URBAN ASTRONOMY

STARGAZING FROM TOWNS & SUBURBS

ROBIN SCAGELL

FIREFLY BOOKS

Robin Scagell is a long-serving Vice President of Britain's Society for Popular Astronomy (www.popastro.com). A lifelong stargazer, he has worked as an observer and photographer, and as a journalist has edited a wide range of popular-interest magazines. Robin is the author of several popular astronomy books, and has contributed to many other publications. He has been awarded the Sir Arthur Clarke Award for Space Reporting in recognition of his many appearances on TV and radio talking about astronomy and space.

A FIREFLY BOOK

Published by Firefly Books Ltd. 2014

Copyright © 2014 Robin Scagell

First printing

Publisher Cataloging-in-Publication Data (U.S.)

A CIP record for this title is available from the Library of Congress

Library and Archives Canada Cataloguing in Publication

A CIP record for this title is available from Library and Archives Canada

Published in the United States by
Firefly Books (U.S.) Inc.
P. O. Box 1338, Ellicott Station
Buffalo, New York 14205

Published in Canada by
Firefly Books Ltd.
50 Staples Ave, Unit 1
Richmond Hill, Ontario L4B 0A7

Published in Great Britain by Philip's,
Endeavour House, 189 Shaftesbury Avenue,
London WC2H 8JY
An Hachette UK Company

Printed in Thailand

Title based on *Philip's Observer's Handbook: Astronomy from Towns and Suburbs*, first published in 1994.

Title page:
The North America Nebula (left) and Pelican Nebula in Cygnus, photographed from light-polluted Newport in south Wales by Nick Hart through a 100 mm refractor using an H-alpha filter plus RGB filters.

CONTENTS

PREFACE

In London's Regents Park, a throng of people gather each month after dark for a star party. They call themselves the Baker Street Irregular Astronomers, after the fictional Sherlock Holmes' gang of local informants, the Baker Street Irregulars. Despite observing in a most unpromising location, they manage to observe a wide range of objects, including deep-sky favorites such as galaxies, planetary nebulae and star clusters. While the views they get of such wonders aren't as good as those from dark-sky sites, they prove the point that city observing is not a waste of time.

But this book is not just about observing from the middle of a big city. More amateur astronomers than ever now have to face the problems of light pollution. You may live in the leafy outskirts of a pleasant small town yet still be unable to see the Milky Way. And even well away from the bright lights, light pollution is still evident. You can be in what seems to be a remote country site, with no habitation or artificial lighting visible for miles, but when you try to photograph the Milky Way, for example, the background rapidly turns brown from the lights of cities well beyond your immediate horizon. Even if the sky overhead is black, lower down you see the familiar skyglow.

Despite this, amateur astronomy is still as popular as ever, with new technology helping to overcome what the streetlights have robbed from us. And purely visual observers, armed with no more than a telescope, can still enjoy good views by choosing their moment and their targets. Writing for the first edition of this book in 1994, veteran US observer Leif Robinson recalled his first ever observation as a child using a small telescope propped on a staircase: the double star Albireo, with its contrasting yellow and blue hues. The good news is that this delight still awaits observers wherever they live, as do many more.

This book is as much for relative newcomers to astronomy as it is for those who have been observing for years. I have had to assume that you are familiar with the basics of observing, so if you don't know what an equatorial mount is, or whether first-magnitude stars are brighter than sixth-magnitude ones, then please look in a more general book. My own *Firefly Stargazing with a Telescope* covers this, and more.

In Chapter 4, I cover observations that you can carry out from light-polluted skies that have a practical use, but the fact is that the vast majority of amateur astronomers observe for fun, cherry-picking the objects that are around, rather than engage in programs of work. You don't have to be a dedicated observer in order to call yourself an amateur astronomer, but around 90% of all amateurs probably don't observe regularly or have a particular subject area that they are interested in. If this is you, don't worry – you're in good company, me included.

I hope this book will help you to discover what you can see, and encourage you get out there and look – or maybe regain the enthusiasm that deserted you when the night sky got brighter. The Universe is still out there and waiting to enthrall you.

▲ A view of Orion over the Houses of Parliament in London, UK, taken in December 2013, using a Canon 40D camera at ISO 400 with a 4-second exposure at f/3.5. Moonlight increased the sky brightness.

1 • INTRODUCTION

A clear deep-blue evening sky is a delight, whether you live in the heart of a big city or in the depths of the country. The Sun has set, the first stars are starting to appear high up, and there is the promise of a fine night's observing. This is the time when every astronomer feels the tug of the stars.

By "astronomer," I mean anyone who wants to be one: you do not need to be a professional, or even a committed amateur with an observatory full of the latest equipment. It is possible to be an astronomer at heart without ever thinking of owning a telescope. Anyone who has gazed at the stars and has wished they could put names to what they see has felt the stirrings of astronomy. After all, the word "astronomer" simply means "star namer." It would be quite easy to make up your own names for the pattern of stars – there is an obvious saucepan; that group of stars below it looks just like a cat sitting on a fence; there's a kite, over to the left. But others have already given these patterns names, so it is those that we set out to learn instead: the ancient astronomers saw a bear, a lion and a herdsman in the stars.

It is a strange feeling, as you look up, to realize that those stars glimmering in our modern electric sky with its unnatural pink glow are the same ones that the ancient stargazers looked upon and named. Yet there is a world of difference between what most of us can see and what they could see. If you live in a city or town, or out in the suburbs, the promise offered by that clear blue twilight sky is usually a false one. The wonder of stargazing has been taken from us by the needs of millions of people to drive, to work, or to be entertained at night-time, and to feel safe from nocturnal criminals. The artificial lighting that makes all this possible sends light up into the night sky, producing what is called light pollution. Figure 1.1 shows how light pollution affects our view of the stars, and Figure 1.2 shows how it makes the Earth look from space.

Most suburban skies show only stars brighter than magnitude 3.5 or 4.0, even at the zenith. There are fewer than 500 stars in the entire sky brighter than magnitude 4.0, of which about 250 are above the horizon at any one time (actually slightly fewer in the northern hemisphere, slightly more in the southern hemisphere). Of these 250, many will be so low in the sky that they are either not visible, or are hidden by buildings, trees, or other obstructions along the horizon. So, out of the 9,000 or so stars that are visible to the naked eye in the entire sky under ideal conditions, from the suburbs you can probably see only about 100 at any one time from your home, even down to the very faintest.

▲ *Fig. 1.1 Two photographs of the Orion region showing the effects of light pollution. From Kensington Gardens in London on a very clear night (left), the sky brightness drowns out all but the brightest stars. In a typical UK country site near Shrewsbury, Shropshire (right), many more stars are visible although there is still some light pollution evident.*

While this makes the task of learning the sky very much easier, without the allure of a star-filled night the invitation to explore further is nothing like as powerful. This, I believe, is the main reason why fewer people actually observe today than in former times. With modern techniques, as we shall see, it is still possible to observe a large proportion of the objects you might want to see. And with the advent of imaging devices, amateur astronomers can reach farther into the Universe than ever before, even with the handicap of light-polluted skies. But there is little doubt that a hazy pink sky does not promise as much as a dark, crystal clear one. It is as inviting to astronomers as a murky pool is to swimmers.

The poor skies also play their part in dampening any initial enthusiasm for astronomy. However, even with a cheap pair of binoculars you can locate many celestial objects that are invisible to the naked eye – although, sadly, few people ever make the attempt. The message of this book is that from an urban site you can still see a lot without having to spend a lot. As for the types of object you might wish to observe, such as planets, stars or nebulae, about two-thirds of them can be observed from the city, including most of the spectacular examples. All you really need when confronted with bright night skies is enthusiasm.

Serious astronomy?

Some astronomers are a little like those hardy backpackers you come across in country areas who make you feel guilty for using your car to drive down the lane rather than walk. Just as some backpackers seem to believe that unless you are suffering you might as well not bother, those astronomers would say that astronomy is worth doing only if there is a result at the end of it. Serious observing is the important thing: if you are not making a contribution to knowledge, you are wasting your time.

But you can enjoy the countryside through your car window if that is what you want. Strolling down the lane is fun too, and just because you never go more than half a mile from the car you will not be enjoying it any the less. Similarly, just gazing at the showpieces of the sky is rewarding, even if they are the same ones you saw last week or last month or last year. You should not feel guilty for not pushing forward the frontiers of science.

On the other hand, if you would like to have some sort of goal in mind then there are lines of work you can pursue from the city just as effectively as from the country. Amateurs often like to believe that what they do is scientifically useful, and indeed it can be. There was a time, often called the Golden Age of amateur observing, when amateurs could make genuine discoveries. The eminent Victorian observer William Lassell, for example, discovered several planetary satellites, Triton among them, while making his money as a Liverpool brewer. During the last century, until the Space Age, amateurs played a leading role in monitoring the Moon and planets. Although this first Golden Age is definitely over, work in some traditional amateur fields, such as variable-star and even planetary observing, has not become outmoded despite the advent of improved observing methods.

But for the serious amateur a new Golden Age is just beginning. At one time, many amateurs had to make their own specialized equipment. These days, there is a wider range available off the shelf and at affordable prices than ever before. And with the arrival of electronic imaging, the well-equipped amateur is now in a position to carry out types of measurements, such as astrometry and photometry, that were previously virtually out of reach. Even within the Solar System there are still discoveries to be made. Amateurs spotted a rare new storm on Saturn that eluded the spacecraft Cassini, which was in orbit around the planet, because it happened to be on the far side when the storm erupted. With a comparatively modest outlay, an enterprising amateur can now enter fields that were once the exclusive preserve of professionals.

The cost

What can you see, and what will it cost? Apart from the fun of learning the sky, which costs virtually nothing, you can get a long way with nothing more than a pair of binoculars costing about the same as a

You can, if you wish, spend huge s...
sophistication. A reasonable next ...
reflector of 250–300 mm (10–...
photograph a wider range o...
System bodies, which riv...
about the same as yo...
very important add...
charge-coupled ...
this let you ta...
by conven...
seconds...
costs...
th...

▲ *Fig. 1.2 A composite i...*
the Earth by night from s...
The glare from millions of ...
upward into space, creatir...
skyglow that outlines each continent.
Although the intensity of light generally
indicates population density, there are
exceptions – France, despite its affluent

...brilliant lights in the Sea of Japan
mark the position of squid-fishing
fleets, which use light to attract
their prey.

budget digital camera. These will show you, for example, the more prominent star clusters and nebulae, even from badly light-polluted skies, as well as bright comets. If you want to make useful observations, you can tackle variable stars and even meteors.

The next stage up is to acquire a telescope. Be extremely wary of very small telescopes of the sort offered in mail-order catalogs or by non-astronomical suppliers. Although some of them are quite suitable for the beginner, most are of poor quality, and by the time you find out you have bought a lemon there is little you can do about it. The telescope market now offers a wonderful variety of instruments to suit all pockets, sometimes literally, but I still believe that a good 150 mm (6-inch) reflector, like that shown in Figure 1.3, is very hard to beat for value and usefulness. You can buy a good starter instrument for less than the price of a budget laptop.

With such a telescope you can see considerable detail on the Moon, pick out markings on the bright planets, and get views of some 400–600 deep-sky objects, even from badly light-polluted areas – though admittedly you would get more exciting views from a country area. By spending more on the right mounting, motor drive and other equipment, you can also take impressive photographs that will show few signs of the effects of light pollution.

...ums on telescopes of increasing ...step up might be a well-mounted ...2 inches) aperture, so that you can ... objects, particularly the brighter Solar ...al those taken anywhere. This would cost ...d might spend on a good used car. Another ...tion to the amateur astronomer's armory is the ...device, better known as the CCD. Not only does ...ke amazing images of objects that cannot be recorded ...ional photography, but it also reveals objects within a few ...that are otherwise totally invisible. A typical CCD camera ...more than the computer you need in order to display and store ...e CCD images (not to mention the permanent observatory of some sort that you might then want to keep everything in).

No matter how much you spend, you will never be able to defeat the streetlights completely. The naked-eye spectacle of the Milky Way can now be enjoyed only well away from urban areas, and in some cases only by traveling abroad. Similarly, from the city, deep-sky objects such as clusters and galaxies will rarely match the descriptions you will find in observing guides. Most galaxies are either invisible or merely ghosts of what they are like in truly dark skies. Your own wide-field color photographs of the heavens are also likely to be a disappointment. Chapter 7 deals with how to find those dark skies that will allow you to observe the objects that are otherwise hidden from you.

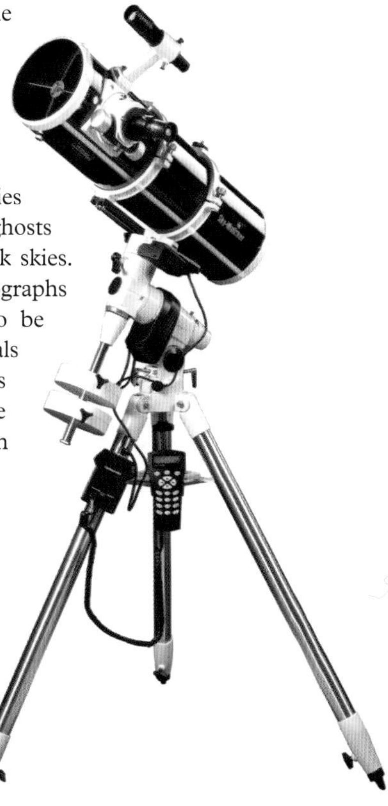

▶ *Fig. 1.3 A 150 mm (6-inch) reflecting telescope on a motorized equatorial mount with Go To that will find objects for you at the touch of a button. You can adapt it for basic long-exposure astrophotography through the telescope using a suitable camera or a CCD.*

2 · KNOW YOUR ENEMY –
THE WEATHER

Astronomers are at the mercy of the weather. The new observer soon finds that it has plenty of tricks up its sleeve. Urban astronomers in particular need to keep a keen eye open, as they have not only the weather to contend with but also the light pollution. The two are linked, as we shall see.

When it comes to the effect of the weather on observing conditions, myths abound. Several of these I aim to correct:

1. Light pollution is the only reason why the city skies are bright.
2. Rain washes the dirt out of the air.
3. Mist creates good seeing.
4. In an unreliable climate or light-polluted area there's no point in doing astronomy.

In the course of this chapter I shall demolish each of these myths in turn, and show how you can improve your observations, even from the city.

The role of water vapor

Light pollution is not the only reason why city skies are so bright. Have you ever wondered what the city lights are shining on? The atmosphere contains more than just air. It carries gases, aerosols, and dust particles from industry, aircraft, wind-blown soil, forest fires, volcanoes, and meteoroids, maybe pollen grains, and, above all, water vapor. Except over certain areas, the effects on sky clarity of all the others are usually trivial compared with that of water vapor.

Water vapor is different from water droplets or steam. It is the gaseous form of water, and it behaves like a gas. Even when the temperature is well below freezing there can still be water vapor in the atmosphere, though warm air can hold very much more water vapor than cold air can. That is why washing on the line dries best on a warm day. Boiling is a quick way of converting water to vapor, but water will still evaporate at lower temperatures – a wet road does not need to be heated to the boiling point of water in order for it to dry out.

The amount of water vapor in the atmosphere is measured by the humidity. (More specifically, it is measured by the *relative humidity*, the ratio of the actual water vapor content to the maximum possible content.) Although water vapor is transparent, the higher the humidity the greater the likelihood that it will condense into fine droplets, causing haze. From the astronomer's point of view, the humidity is

◀ Fig. 2.1 (top) A clear day in Britain, with the sky milky as a result of water vapor in the atmosphere. This is also common throughout much of Europe and the US. (bottom) The sky over the Mills Cross radio telescope at Molonglo near Canberra in Australia, however, often contains less water vapor and is a much deeper blue.

an important quantity: high humidity means milky skies during the day and poor clarity at night. Britain, for example, is not generally thought of as having a humid climate, but humid does not necessarily mean warm. It can be cold and humid – as in a November fog, when the air is saturated with water and the relative humidity is 100%.

In the daytime the difference that water vapor makes to the atmosphere is quite evident. You can generally tell by the color of the sky, or by the quality of the air itself as viewed against distant objects, what is contributing to the haziness. Water vapor turns the sky milky. When it is very humid the sky is pale blue even at the zenith, and near the horizon the sky is virtually white. This is quite different to the effect of industrial pollution, which often lends the sky a reddish or brownish tinge, and to heat haze, caused by dry particles, which is a bluish-gray.

Water vapor is as much of a problem to the urban astronomer as streetlights are. If the skies were crystal clear, free from all water vapor and dirt, those faint objects that we strain to see would be much easier to observe. It is because the water vapor in the atmosphere reflects light shining up from the ground back down again that the night sky takes on its orange or pink tinge. Not only does the water vapor reflect artificial light in this way, it also dims the light from the distant objects we want to observe. The two effects combine to prevent us from seeing anything but bright stars.

The amount of water vapor in the atmosphere varies considerably from place to place. In Britain and along the US East Coast, for example, the sky is usually milky. Very rarely do observers there get to enjoy the deep-blue skies characteristic of such places as Australia and

Arizona. As Figure 2.1 shows, even near Australia's capital city the skies are usually much bluer than British skies. An Australian woman who visited Britain for the first time commented that the sky seemed "much closer than in Australia." Perhaps its milky appearance made it seem nearer, like clouds, than the distant deep blue she was used to.

The reason why Britain in particular has milky skies is its geographical location. Not only is Britain an island at the edge of the Atlantic, it is also right in the path of atmospheric *depressions*, or *troughs* – regions of low pressure – which trundle across the ocean, picking up moisture as they go. As tennis matches at Wimbledon are interrupted by rain yet again, the citizens of Paris, a little over 300 km (200 miles) to the south, may be enjoying unbroken sunshine. The tracks of the depressions vary from year to year, sometimes moving farther south so that Paris shares the same gray skies. In other years the depressions favor Iceland, and Britain enjoys a hot summer.

We now know that these variations are the result of changes in the jet stream, a high-altitude wind system that snakes around the middle to northern latitudes of the northern hemisphere. It is linked to other global variations that include El Niño, the temperature variation of the Pacific that also has wide effects. Just how all these changes are connected is still a matter for study and indeed hot debate.

Such weather systems are by no means restricted to Britain and the US East Coast: they affect most continental land masses. Yet a depression is not necessarily bad news for the observer. It may well bring rain, but a passing weather system can bring clear skies as well – really clear skies, not those hazy ones that are so bad for urban astronomy.

A depression forms at temperate latitudes when a bubble of warm, moist tropical air pushes into an area of cold, dry polar air, as depicted in Figure 2.2. The air pressure at the edge of the bubble is low, and the bubble forms into a spike with its pointed end in the center of the low-pressure area. The winds within a depression always blow counterclockwise in the northern hemisphere, clockwise in the southern hemisphere. The leading edge of the spike of warm air is known as a *warm front*, and the leading edge of the cold air as a *cold front*. On a weather map, fronts are indicated by the familiar semicircular and triangular symbols.

Clouds form along fronts, and the passage of a front, either warm or cold, is often marked by rain. After a cold front has passed by, the weather turns fresher and brighter as the cold, drier polar air washes over the land. This polar air is usually a great deal clearer than the warm humid air which preceded it. The skies take on a bluer appearance by day, and are much darker by night. The haze has gone. You will then hear recited Myth Number 2: "The rain has washed the air clean."

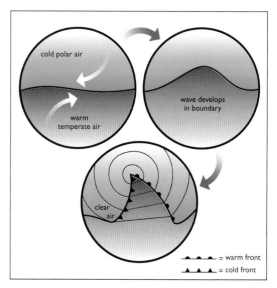

cold polar air

warm temperate air

wave develops in boundary

clear air

◄ Fig. 2.2 A frontal system begins where mild temperate air meets cold polar air. The two air masses have different densities, so they do not mix but instead slide past each other. A ripple in the layer between them can develop into a larger feature, which eventually becomes a fully fledged depression. The triangular symbols indicate a cold front, the semicircles a warm one.

= warm front

= cold front

Although it looks that way, that is not really what has happened. No doubt the rain has washed any dirt particles out of the air mass that was over you at the time, but, as long as the wind is blowing, the air that surrounds you at midday is miles away by 1 p.m. Clear skies can follow rain because they belong to a different air mass which has moved in to replace the one that was there before.

To the urban astronomer this clear air is just the ticket. For a few hours at least there is the chance to observe those faint objects that have been eluding you. Your main problem may well be some cumulus clouds scudding along, making observing rather frustrating. Even so, some of the best views of the sky can be had in gale-force winds. In between the clouds the skies are very transparent, with almost no suspended dust or water vapor to reflect the streetlights. Observer and telescope must be well protected from the elements, both for the sake of comfort and to prevent the telescope from wind-induced vibration. The scale of this vibration is small, but if you are photographing a celestial object it is crucial. Telescopes and cameras can vibrate imperceptibly. You will probably not suspect that anything is wrong until your photographs come out unaccountably blurred.

Making the most of the occasions on which a cold front has just passed through is an important part of beating the streetlights. Unfortunately, such opportunities can be rather rare, though it does depend very much on where you live. Not only must the front have passed through at the right time, but the skies behind it must be free from cloud – rarely do weather fronts behave as they do in textbooks. And, of course, you must be out there to make the most of it.

Cold fronts associated with depressions are not the only source of cold polar air. High pressure can also do the same. In this case, the air circulates clockwise in the northern hemisphere. If the system is large enough, the eastern edge of a high-pressure system can also bring cold, clean polar air down to lower latitudes. But the middle of a high-pressure area, while regarded as bringing settled, fine weather, is often poor for astronomy because of the buildup of haze. Amateur astronomers quickly learn that the appearance of a sunny symbol on weather maps doesn't always promise a lovely clear evening good for astronomy. On a clear but hazy night following a hot, sunny day you can sometimes see hardly any stars.

In some areas, winter means snow. While cold air streams can be low in water vapor, snow will have an adverse effect on sky brightness – light from streetlights which would otherwise be absorbed by the dark ground is instead reflected up into the sky with high efficiency. This can counteract the advantages of the clear air. If there happens to be an aurora as well, you can probably forget deep-sky observing!

When water vapor is welcome

There are actually occasions when water vapor is welcomed by astronomers. Myth Number 3 states that "Mist creates good seeing." By *seeing*, astronomers mean the steadiness of the air through which they are observing (though sometimes the term is wrongly applied to atmospheric clarity, or *transparency*). The state of the seeing is particularly evident when you are looking at the Moon or planets. Only too often the limb (edge) of the Moon, for example, seems to be in constant motion. It shimmers and shakes as if the whole lunar surface is in turmoil. Your view of a lunar feature is spoiled as the image dances around, blurring, growing, shrinking, doubling, as you can see in Figure 2.3. Details on a planet such as Jupiter are frustratingly difficult to pick out; just as soon as you see them they are gone again. The annoying thing is that the night looked so invitingly clear, with the stars twinkling beautifully. When the seeing is good the planets are rock steady, their images showing hardly a movement.

A widely used practical scale of seeing was devised by the French astronomer Eugène Antoniadi, and is named after him. It runs from I, denoting absolutely perfect, without a quiver, to V, for terrible. As far as I am aware, there is no conversion between this subjective scale, used largely by amateurs for planetary observation, and the objective scale of seeing for point sources observed with large profes-

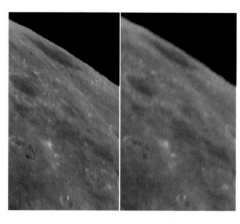

◀ *Fig. 2.3 What astronomers call "seeing" is atmospheric turbulence that varies very rapidly. Here are two consecutive frames from a video sequence of part of the Moon imaged using a 200 mm telescope, showing the variation in seeing quality over ¹/₁₀ second.*

sional instruments, which is derived from image size and measured in arc seconds.

Professional astronomers have searched the world for sites where the seeing is as close to perfect as possible, the goal being 1 arc second or below. At a small number of sites, seeing as good as 0.25 arc seconds has been achieved. (The seeing does also depend on whether you use a small or a large telescope.)

Although good seeing is particularly important when observing details on the Moon and planets, it is much less of a factor when looking for deep-sky objects: good transparency is then what really matters. However, when searching for difficult, faint objects, such as remote galaxies or the faintest stars, poor seeing can still be a hindrance.

Generally speaking, however, the two sorts of observing are mutually exclusive. Clear, sparkling nights are often accompanied by bad seeing, while nights when the planets beam steadily down without so much as a shiver are often rather misty. But back to the myth that mist improves the seeing. What is really happening is that still air, which gives good seeing, also encourages mist to form as the ground radiates its heat away into space and water vapor condenses out of the air, forming dew where it comes into contact with cold surfaces. (The exposed lenses and mirrors of your telescope are ideal surfaces for the formation of dew – another problem for the amateur astronomer.) So, although planetary observers often regard a little mist with benevolence, it is really just a sign that the seeing is likely to be good, rather than the cause of good seeing, though the water vapor does help to even out temperature differences.

The steady air which gives rise to good seeing is most likely to come during a spell of high atmospheric pressure – sometimes called a *ridge*. The daytime skies may become hazy, and the wind may drop. Sometimes, especially in winter, these conditions can also be accompanied by a dreary blanket of cloud, or by bitterly cold, frosty weather. Mel Bartels of Eugene, Oregon, describes how conditions in his area can change:

> "Local winds do little harm to the image, though they shake poorly mounted 'scopes. High-level winds destroy the image when a front (usually a cold front) first passes through. A typical progression here in the NW goes like this: cloudy for a while, then just after a front passes through, very transparent but with poor seeing, then the following night, good transparency and good seeing, followed by a night or two of gradual transparency loss, then come the clouds to start the cycle over again. In the last half of summer and first half of fall, we typically have clear skies for just about the entire time, with transparency and seeing at their best for the first several nights after things clear ...."

This progression, however, is rarely experienced by British observers, who often have to put up with cloud, followed by rain, followed by cloud again.

It therefore pays to keep a close eye on the evening weather forecast. Unfortunately, some TV forecasts avoid showing a proper weather map with standard symbols, or even giving any indication of what is really happening. You may have to work out what is going on from the presenter's chatter.

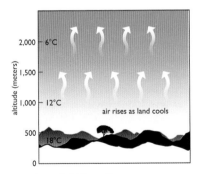
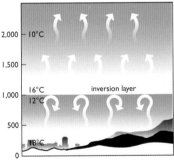

▲ Fig. 2.4 Normally, air temperature decreases with altitude at an average rate of 0.6°C per 100 meters (1°F per 300 feet), as on the left. But where there is a temperature inversion, warm air overlies colder air, and over a small altitude range there is an increase of temperature with height. The pollution-laden air near the surface is trapped below the inversion layer, above which the air is stable and clear. Inversion layers usually occur at heights of 1.5 km (1 mile) or more, but occasionally they reach almost to ground level.

◀ Fig. 2.5 This 1970s view from the Pic du Midi, at an altitude of 2,800 meters (9,200 feet) in the French Pyrenees, shows a temperature inversion marked by a flat layer of haze. The telescope dome in the foreground was set apart from the main observatory building in order to avoid local seeing effects caused by the concrete structure.

Good seeing is also found above what is termed an *inversion layer* (see Figure 2.4). The normal temperature profile for the atmosphere is that it gets cooler with increasing altitude, and on a large scale this always applies. But if the lower levels are cooled by some other agency, such as a cold sea current, the reverse may occur. This is what happens on the eastern side of the Atlantic and Pacific Oceans: the west coasts of Europe, Africa, and the Americas are washed by currents which are unusually cool, having brought water from polar regions. On a small scale the normal temperature profile of the atmosphere is inverted, a layer of cold air forming beneath a warm layer. The two layers are of different densities, and stay as separate as oil and water. The air in the lower layer is unable to rise, causing the notorious fogs and pollution of such areas as San Francisco, California.

Above this layer, however, the air is both stable and clear. Temperature inversions are not to be encountered everywhere, as they are normally some way above sea level, so observers in London or New York are unlikely to benefit from them. But if you are fortunate enough to encounter one, you will find conditions ideal for astronomy. Figure 2.5 shows an inversion layer as seen from the Pic du Midi Observatory in the French Pyrenees.

The search for good seeing

Whether you live in the town or the country, what the seeing is like will determine how successfully you can observe the planets. So the question is, does the town give any worse seeing conditions than the country?

First, there is high-level seeing and local seeing. *High-level seeing* is determined by weather systems, such as the passage of fronts and the presence of jet streams, and affects both urban and rural areas equally.

You can often see the turbulence that causes poor high-level seeing by defocusing the telescope while looking at a bright star. Turn the focuser so that the eyepiece comes out toward your eye, and you are focusing on a point closer than the star. The star will expand to a disk, against which you can see what appears to be a rapid flow of material, rather like looking at the surface of a turbulent stream. You are actually focusing on a point only a few hundred meters from the telescope, though the disturbances you see may be farther away than that.

In the early 1970s I was stationed for a while at the Pic du Midi. Astronomers there experimented with focusing on this turbulent layer in order to estimate its height. Unfortunately, it proved too difficult to focus on accurately, even though they were using telescopes with very long focal length, giving much finer discrimination of focus position.

Local seeing is caused by disturbances in and around the telescope itself. These begin within the tube, more so with reflectors, which are usually open at the top end, than with reflectors, in which the air in the tube is sealed in. People often complain of bad seeing when in fact they have simply not allowed the telescope sufficient time to cool down. If the air in the tube has not cooled to the temperature of the night air, turbulent currents will be set up, which will ruin the seeing. Ways of improving your telescope are dealt with in more detail in Chapter 5.

Even refractors are not immune from this. I have seen an expensive refractor take at least an hour to settle down after having been brought out from the warm. The urban astronomer's telescope may well be kept indoors, so this is an additional source of problems. The solution is to put the telescope outside a good hour or so before observing, though with all the covers in place to prevent its optics from dewing up.

Moving outward, the next common cause of bad seeing is air currents generated by heat from the observer. This happens particularly with reflectors, where the observer peering into the eyepiece is very close to the telescope's line of sight. The remedy is simple – position yourself so that you are not beneath the tube, and make sure you don't breathe into the line of sight. The ideal observer would not breathe at all!

This effect is worse if you have an observatory. It is natural to invite people round to look through the telescope. Those who are not observing usually peer out of the dome opening, right under the top of the tube. The heat generated by each person pours out of the opening, giving lousy seeing which miraculously improves as soon as they have all gone home.

Beyond your telescope there are many other causes of bad local seeing. Anything which collects heat during the day is a potential problem. Not only its color but its thermal properties are also significant.

Dark objects collect more heat than light ones, while good insulators (such as concrete) will lose heat more slowly than good conductors (such as metal). The city is full of concrete, and holds its heat well. Temperatures are usually a few degrees warmer in the center of town compared with even the suburbs.

There are some obvious features to avoid. A neighbor's chimney is the first thing that comes to mind, but these days fewer people light fires and many chimneys are unused. Homes are warmed instead by central heating, which is much more sneaky. The vents, along with those of air-conditioning units, are generally on a side wall, and can shoot a jet of hot, moist air several feet out from the building, just where you least expect it. The boiler probably switches on and off every so often, leaving you puzzled about what is going on. If the boiler provides hot water, it can be operating even on warm summer evenings.

Uninsulated roofs and tarred driveways are notorious harborers of heat. Many urban observers find that the seeing deteriorates if the object they are viewing moves over a roof. Grassy areas, on the other hand, are pretty innocuous, and an expanse of water can be ideal. But country-dwellers need not feel completely smug: broadleaved trees, when in leaf, have been accused of releasing parcels of heat throughout the night. Taking this into account, you might try moving your telescope around, if it is portable enough, in order to get better seeing. It is possible that you can get an improvement in seeing simply by avoiding a local source of stored heat.

At major observatories, conifers may be planted around the domes where possible. They keep the ground cool, and – unlike broadleaved trees – are said to release relatively quickly any heat they have stored from the daytime. Observatory domes are usually painted white to keep them cool during the day (Figure 2.6). The aim is to maintain their interiors at night-time temperatures so that no time is wasted cooling everything down after opening the dome.

◀ Fig. 2.6 The dome of the William Herschel Telescope on the island of La Palma in the Canary Islands, like other domes, is painted white to minimize daytime heating. The approach roads may even be covered with white gravel in an attempt to eliminate the effects on seeing of the dark asphalt, which at night gives off heat it has absorbed during the day.

It may sound as if observing from the town is pointless, with all the sources of bad seeing around. Many people will tell you that city seeing is terrible, even without having experienced it. Yet urban observers often manage to make perfectly good observations. Planetary observer John Murray used to live in Luton, Bedfordshire, just a short distance from where the legendary telescope maker Horace Dall had his observatory overlooking the urban sprawl. John found that the seeing when viewing over the town could be excellent. His line of sight was a hundred meters or so over the rooftops, so that individual local sources of bad seeing were evened out. Planetary imager Eric Ng lives in a high-rise flat in urban, light-polluted Hong Kong. This has to be one of the most unpromising locations on the planet for astronomy, yet his images are world class. Others who have observed from both town and country tell similar stories and confirm that in practice they notice little difference when viewing the planets.

David Frydman has observed from builtup areas in London and Helsinki for many years. He thinks that city seeing is actually better than in the country. As long as the line of sight is well above obvious disturbing influences, it can be exceptionally good. He also finds good seeing when the night-time and daytime temperatures are very similar, and as a consequence there is little escape of heat from the ground. In Helsinki this happens in the fall, when the sea temperature of the nearby Baltic and the air temperature are very similar, about 16°C (61°F). Mel Bartels confirms this from his own experience in Oregon.

Although the planets are bright enough to cut through murky skies, your eyes do need a certain brightness level in order to make out low-contrast detail. Jupiter is bright enough for you to make successful observations, even through misty skies. It is an ideal object for observing from the city, as indeed is the Moon. Saturn and Mars, however, are usually dimmer and have lower-contrast features. Poor transparency can therefore wipe out the features you are looking for, even if the seeing is perfect.

Though you can sometimes get a good view of a planet even if it is fairly low in the sky, if your taste is for deep-sky objects David Frydman recommends that you concentrate your efforts on objects that are above about 50° altitude. Below that level there is a marked cutoff point as the transparency of the atmosphere decreases and the brightness of the sky increases, as Figure 2.7 shows. I suspect that many city-dwelling astronomers give up deep-sky observing simply because they have become disillusioned by looking for objects when they are too low in the sky.

Seeing can improve during the night, as temperatures settle down and the ground loses its heat. The dedicated observer who prefers to view the morning rather than the evening sky can be rewarded with much better conditions. Optical worker Jim Hysom remembers the perfect seeing he encountered while serving with the British Army in Gibraltar. He was on guard duty, equipped with a telescope. He found that just before dawn he could obtain pin-sharp images of the Moroccan coastline, about 20 km (12 miles) away. But a few minutes before sunrise the seeing would suddenly change to the familiar daytime hazy blur. The cause of this, he believes, is that infrared radiation from the Sun, which is refracted more by the atmosphere than its visible light output, "rises" first and warms the air before the Sun itself rises.

Solar observers have more to contend with than their night-time counterparts, since the source of the heat that spoils the seeing is necessarily present. Local seeing is likely to have a greater effect than at night, since surfaces are being heated all the time, even on a cold winter's day. Solar observer Geoff Elston says that even from an observing site in Clapham, south London, the seeing for solar observing can be good, mostly in the earlier part of the day.

Big Bear Solar Observatory in California is set in the middle of a lake, which steadies the local seeing. At other solar observatories, mirrors high above the ground collect the Sun's light and beam it down to where it is needed.

Interestingly, the seeing also depends on your telescope's aperture. The French astronomer Audouin Dollfus, who observed extensively from a wide variety of sites (including high-altitude balloons), offered the following explanation. Particularly at night, when the ground is cooling and the air above may be warmer, the lower atmosphere consists of layers of air of different temperatures. As air is a poor conductor of heat, these layers tend to remain separate, with well-defined boundaries between them. The layers and their boundaries undulate, perhaps even developing eddies like the swirls in a flowing stream.

As a result, the incoming wavefront of light from a star, which should in theory be smooth, is distorted on a scale which varies from 0.1 to 1 meter (4 to 40 inches). These

◀ Fig. 2.7 The visibility of faint objects depends on their altitude. This fisheye view shows that above about 50° the sky becomes considerably darker.

deviations in the wavefront are often termed *cells* of air. (They are not necessarily the same as the air pockets and other forms of turbulence encountered by the pilots of light aircraft.) The beam of light from a star to your eye is being distorted by these cells, which are constantly in motion. It is this movement which gives rise to twinkling. You can get a quick idea of how good the seeing is by assessing the amount of twinkling: the less twinkling, the better the seeing.

When night and day temperatures are similar, the whole lower atmosphere tends to settle down and the layers disappear. This is what happens where there is a ridge of high pressure. Air saturated with moisture (giving a relative humidity of 100%), as it is when there is a mist, conducts heat more easily and will thus lose any layered structure it may have had. As mentioned on page 15, the seeing improves under such circumstances.

Amateur astronomers often say they get better seeing with a small telescope than with a large one. The reason now becomes clear: if the seeing cells are about 100 mm across, then a telescope of roughly this aperture will be looking through little more than one cell at a time. The image will be sharp, and although it may move around a bit, it will still be easy to observe. A larger telescope at the same site will be looking through two or more cells at the same time, with inevitable blurring. This is why you sometimes see double or triple images. I suspect that this effect is one of the reasons for the claimed superiority of refractors over reflectors, for refractors are generally of smaller aperture.

Observers know that there are actually many different types of seeing. Sometimes everything shimmers or wriggles, but everything looks quite sharp. Objects remain in more or less the same place. At other times, the image looks as if it is on a rubber sheet that is being pulled in and out in different places, and a crater on the Moon or the outline of a planet constantly changes shape and moves around from one second to the next. Occasionally, everything seems to explode and becomes blurred for some time.

If observing from the city is deemed to be a waste of time, then observing from a high-rise building or an upper-story window is surely total folly. There are several factors which conspire to ruin your view. First, the building will be giving off heat it absorbed in the daytime, setting up an invisible wall of bad seeing around it. Second, observing from an open window means letting all the heat in the room flow out of the window, passing around the telescope, which is also bound to ruin the seeing. And third, you just cannot get a steady enough base: an upper floor of a building is far too susceptible to vibrations compared with terra firma.

That, at least, is what some people say, but others have proved them wrong. The balcony of a high-rise may not be the best place to observe from, but it does have its advantages. Not the least of these is the fact that it is accessible. I have always said that the 150 mm (6-inch) telescope set up at home is worth far more than the 400 mm (16-inch) half-an-hour's drive away. If you can get ready within minutes, you are much more likely to observe than if you have to spend ages getting everything into place. If the building has concrete floors, you should have a steady enough base. Your main worry will probably be the risk of your precious equipment toppling over the balcony.

Although observing through an open window is notoriously bad for the seeing, all amateur astronomers have done this at some time or other, and the results are not always as bad as some would have you believe. It is even possible to carry out useful work from indoors, as detailed in Chapter 8. And if your options are to observe through a window, or go outside and risk being mugged, I know which I'd choose.

One big problem with a high-rise is likely to be turbulence around the building. This, however, will vary with the wind direction. With experience you may find that some winds are more favorable than others. It is probably better to be facing into the wind than in the lee of the building, though the effects may be masked by the fact that different air masses come from different directions anyway. North winds, for example, usually bring different weather from south winds, wherever you live. In fact, it is possible that under the right circumstances the seeing from a high-rise could be very good, since you are above local sources of bad seeing. Professional astronomers, too, have discovered that good seeing is to be had high above the local ground level, and accordingly locate their telescopes well above the ground. The aim is to be in a smooth air flow, with no local eddies.

Even mountain-top observatories have problems with turbulence. At the Pic du Midi, for example, the mountain is the first really high ground that northerly winds encounter. The air mass rushes up the slope and swirls around right over the observatory, creating bad seeing. Winds from the southeast round to the west, however, meet the higher mountains of the main Pyrenees chain first. They dump their precipitation on the mountains, leaving a comparatively tame air mass to encounter the Pic itself, which is in their lee (though a long way downwind).

When it was decided to find a site for a major observatory within easy reach of European astronomers, many sites were tested and a small army of young people was sent out with telescopes and tents to observe the seeing from a number of locations. Having tested inland sites, and the islands of Madeira, Tenerife and La Palma, the latter was eventually chosen. It rises steeply from the Atlantic to a height of about 2,400 meters (8,000 feet), so it often projects into a stable air flow above an inversion layer.

Such effects are not restricted to major mountain chains or steep islands. Many cities are in the "rain shadow" of a range of hills, and it is even possible that city buildings can act as a barrier, if not to rain then at least to turbulence, in the same way.

How bad is the weather?

There are some lucky amateur astronomers who live in places where the skies are often clear and the seeing is often good. But the rest of us have to put up with weather that varies from the mediocre to the atrocious. British astronomers, and those in parts of the US East Coast, feel particularly hard done by, with clear skies between only 20% and 33% of the time. Even when it is clear, the weather seems to know just when you are about to observe. No sooner have you got your coat on and set up the telescope than cloud appears just where you were going to look. You wait a while, then give up. Ten minutes after you pack up, the sky is clear again.

Despite this, some great astronomical discoveries have been made from Britain. British amateurs today have to put up with much the same weather as, for example, did William Herschel in the 18th century. During one night's observing, it became so cold that the ink froze in his inkwell. Admittedly, he did not have light pollution to contend with as well, but my point is that there is enough clear weather even in the worst climate to allow the keen amateur some observing time. The main problem is flexibility. In a poor climate you cannot reserve Thursdays for astronomy, or decide to have a Saturday night star party – you have to grab any opportunity when it presents itself. The difficulties are really social rather than astronomical. So Myth Number 4, that in an unreliable climate it is not worth observing, simply reduces to a question of priorities.

By getting to know the seasonal weather patterns in your area, you can begin to overcome some of its worst effects. Keep an eye on the weather forecasts, and learn what sort of weather gives you the best conditions. You may have to adapt your observing methods to the weather. Rather than wait for perfect conditions for wide-field deep-sky photography, which requires continuous clear skies, for example, try planetary photography instead, or aim for star clusters that can be recorded with shorter exposures through the gaps in the cloud. Either that, or buy a CCD (see Chapter 6).

3 • KNOW YOUR ENEMY – THE STREETLIGHTS

When I was young, the bane of my astronomical life was the streetlight across the road from my back garden in suburban Middlesex. It had a tungsten bulb of maybe 100 watts, and went out at about midnight. I remember looking out of the window at about 3 a.m. to see the dark skies, though my enthusiasm did not stretch to going outside to observe. The evening skies were then still tolerably dark, allowing me to see plenty of objects.

I also remember an Easter trip to the small seaside resort of Dawlish Warren in Devon at about the same time, in the early 1960s. One night it was cloudy, but I thought there might be some clear patches. I made my way down to the shore, but as soon as I moved away from the lit footpath it became so dark that I could not find my way. Today there would be no difficulty in walking around on a cloudy night – the glare from nearby Exeter and other towns, reflected from the clouds, would provide ample illumination. Streetlighting, and other forms of outdoor lighting, have spread far and wide over the past 50 years.

Streetlighting began longer ago than might be imagined. As early as 1405, the Aldermen of the City of London had to ensure that every house alongside a road displayed a lit candle in the window from dusk until 9 p.m. Four centuries later, one of the first uses of town gas (a byproduct of coal, now superseded by natural gas) was to provide streetlighting. In 1765 a colliery manager by the name of Spedding suggested that the streets of nearby Whitehaven in the north of England should be illuminated by gas lights similar to those he had installed in his offices, which used gas from the mine. His idea was turned down. In the end, it was the streets of London that were illuminated by the first public gas lighting, starting in 1807 when lights were installed in Pall Mall by Frederick Winsor.

From then on, gas lighting spread wherever gas companies were set up. Towns became brighter. In his famous novel *Far From The Madding Crowd*, written in 1873, Thomas Hardy refers to the glow from the nearby town of Casterbridge, which in reality was Dorchester, about 5 km (3 miles) from the cottage where Hardy spent his early years:

> A heavy unbroken crust of cloud stretched across the sky, shutting out every speck of heaven; and a distant halo which hung over the town of Casterbridge was visible against the black concave, the luminosity appearing the brighter by its great contrast with the circumscribing darkness.

When within a mile or so of the town, Hardy's character Fanny Robin can see her way "by the aid of the Casterbridge aurora…."

Such lighting must have been much dimmer than present-day urban lighting. Its original aim was to help citizens find their way around without walking into potholes, puddles, and other undesirable features of the thoroughfares and byways of the time. A gas lamp every hundred yards or so was adequate for this purpose. These days, most of us rarely find ourselves out on a truly dark, cloudy night where there is not a glimmer of light from earth or sky. Few non-astronomers keep track of the phases of the Moon, for example, yet to our forebears they were important. In the 18th century a group of eminent men calling themselves the Lunar Society met regularly in Birmingham. They chose the name not because they were particularly keen on the Moon, but because they met each month around the time when it was full, making it easier to get to the meetings.

Eventually, gas streetlights were replaced by electric ones, generally of similar brightness to the gas lamps. Some gas lamps remain even in Central London – in 2013 I found one just yards from Regent Street, unnoticed by the busy shoppers – and you might find them in other towns and villages outside London, though probably they are being replaced all the time.

During World War II, blackouts were strictly enforced in Britain and other countries to avoid giving clues to enemy bombers. The population again had to learn how the Moon's phases vary, in some cases because its light meant that an air raid was more than likely. Many Londoners remember seeing the Milky Way for the first time during the blackout, and quite a few became interested in astronomy as a result. In California, Walter Baade found that he could push the famous 100-inch (2.5-meter) telescope on Mount Wilson, overlooking Los Angeles, to its limits. Since the end of the war, however, streetlighting has increased in brilliance virtually everywhere. Astronomers wish the blackout could return – but without the attendant risk of air raids.

Although streetlights are by no means the only source of light pollution, with floodlights being equally if not more important, by and large the same types of light source are used whatever their purpose.

Types of light

Astronomers have by necessity become very aware of the different types of light source that they see around them. When a new type of lighting appears they are the first to notice, and to try to work out how it will affect their observations. So here is a guide to get you up to speed with the various types and the light that they produce, presented roughly in chronological order of their development.

Studying the spectrum

An important feature of any light source is the spectrum of the light it emits – that is, the distribution of its light across various wavelengths. For astronomers it is crucial to know the spectrum for each source of unwanted light so that some counteractive measure can be taken.

Looking at the spectrum of white light, as produced by a prism or diffraction grating, for example, we see the familiar rainbow of color. What we do not see is that the rainbow extends on either side into "invisible" radiation. Beyond the red end is infrared (which we can feel as heat), and beyond the blue end is ultraviolet (which tans our bodies and can have other, more serious, effects on our health if we get too much exposure to it). The word "radiation" covers anything which radiates, from plain old yellow light and ultraviolet, which are both types of electromagnetic radiation, to particles such as those given off by radioactive substances.

What we see as white light is really the visual impression of all the individual colors seen together. To be more accurate, white light is radiation from a body at a particular temperature – for example the Sun, whose visible surface is at a temperature of around 5,500°C (10,000°F). A block of metal heated to this temperature would give out virtually the same proportions of colors – there are additional absorptions in the solar spectrum – and the same amounts of infrared, ultraviolet, and other forms of radiation.

The feature that distinguishes the different types of electromagnetic radiation is their wavelength. Traditionally, wavelength was measured in angstroms (symbol Å), on which scale the wavelength of yellow light is around 5,000 Å. Today, it is measured in nanometers (nm), 1 nanometer being 10 angstroms. So yellow light has a wavelength of 500 nm, or 500 billionths of a meter; red light has a wavelength of around 650 nm, and blue light around 400 nm. With sunlight, starlight, or an older domestic light bulb, the light consists of all colors, though there is rather less blue and red light than there is yellow. They have what is called a *continuous spectrum*.

Incandescent lighting

The original and familiar type of electric light, the sort we used throughout the 20th century until low-energy bulbs came into use, is the tungsten incandescent bulb. It consists simply of a tungsten wire filament inside a glass vessel from which all the oxygen has been extracted and replaced with an inert gas. The gas plays no part in the light-producing process, but prevents the filament from burning or evaporating as the current passes through it, heating it to incandescence. Although we use the white light it gives off to see by, most of the energy – up to 98% – is converted to heat rather than light. This really is wasteful, and it is for this reason that energy-saving light bulbs and tubes have come on to the market and incandescent bulbs are getting harder to find.

A variant on the tungsten bulb is the tungsten halogen lamp, in which a halogen gas such as iodine is added to the inert gas in the bulb. This prolongs the life of the filament and allows it to be run at higher temperatures, giving a whiter light. However, neither of these types of bulb are generally used for streetlights though they can be used for illuminating shop signs, domestic lighting and floodlamps, including small security lights.

In Chapter 6, I deal with ways of filtering out the glare from streetlights, but since the spectrum of a tungsten lamp is virtually the same as that of a star, tungsten lights are a problem. Overall, they contribute little to the upward-directed light pollution that we have to deal with, though on a local level they can be very irritating in the form of security lights (now often disparagingly referred to by astronomers as "insecurity lights"). These are available in do-it-yourself stores very cheaply, and are often installed so as to shine light over a wide area, including neighbors' properties. Even those controlled by PIR (passive infrared) sensors can be a menace to observing, as they can be triggered by cats and other animals, and even moving vegetation on windy nights.

The spectrum of an incandescent bulb depends on its temperature (Figure 3.1), and in fact the spectra of light bulbs are almost exactly the same as those of any glowing body at the same temperature, even a candle or a star. Low-wattage light bulbs are around 2,700 K, where K is the Kelvin, the scientific measurement of temperature, which is the same as degrees Celsius plus 273 degrees to relate it to absolute zero. A flame is lower still, about 1,800 K. High-wattage incandescent bulbs, as used in security lamps, are about 3,000 K, which overlaps with the temperature of what astronomers call red stars. Very hot

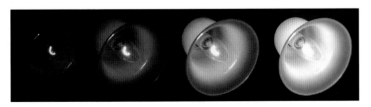

▲ Fig. 3.1 As an incandescent light bulb heats up it changes color from dull red to white. These colors are similar to those of any other body at the same temperature, whether a metal or a star.

objects, and stars, are bluer than daylight. The hottest light source we regularly encounter is lightning, which can reach 30,000 K, as hot as the hottest stars.

A typical incandescent bulb has a brightness peak in the yellow part of the spectrum, but still contains a rainbow of all the other colors, including blue, known as a continuous spectrum (see also page 28). As it gets hotter, the proportion of blue increases. But our eyes adapt to the color of our surroundings, so we see their light as white when they are the only light source. We've all noticed that bulbs that appear pure white at night actually look yellow or even orange when switched on in daylight or from outside at dusk. Cameras set to daylight color balance bring this color out very strongly.

Because the distribution of light from incandescent bulbs matches that from stars very closely, there is little that can be done to filter it out. More about this in Chapter 6.

Mercury lighting

Although many small towns were once lit by tungsten lamps, like the one shown in Figure 3.2, these lamps were found to be very wasteful when brighter lighting came to be needed. Lighting engineers therefore turned to other sources of light. An early success was the mercury-vapor lamp, in which a high electric current is passed through mercury vapor, which then gives off an intense bluish-white light. The result is a bright lamp, well suited to lighting streets. In the late 1950s the old tungsten lamps in the road alongside my garden were converted to mercury lamps. Fortunately, they still had a time-switch which turned them off at midnight.

Mercury streetlights have a different type of spectrum from tungsten bulbs. Instead of a continuous spectrum, the light is concentrated into very narrow lines of single colors – an *emission line spectrum*. There is a strong blue line, a green one, a yellow one, and a rather faint red one. This means that it is possible

◀ Fig. 3.2 A picturesque streetlight from the early 20th century. It has a simple tungsten bulb and uses mirror segments to reflect the light sideways so as to give a wider spread.

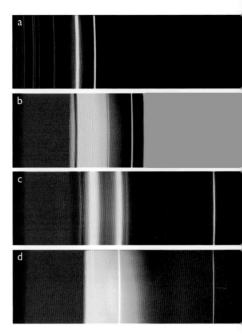

▶ *Fig. 3.3 Spectra of streetlights. (a) Low-pressure sodium, with a dominant pair of orange-yellow lines, here shown as one. The green line is in fact very weak and contributes very little to the light of the lamp. (b) High-pressure sodium, with a dark band at the main wavelength of low-pressure sodium. (c) Mercury, with enhanced red emission as a result of coating the lamp with phosphors. (d) Fluorescent, with additional phosphor coating inside the tube to increase the blue and red output. These photographs of spectra show more detail at the red end than can be seen by eye, and the color rendering of the photographic system produces fewer individual colors than are visible with the eye.*

to see colors under mercury lighting. For you to see a particular color in an object, it needs to be illuminated by light that contains that color in the first place. A red car appears black if there is no red component in the light that illuminates it, for example. As the red line in mercury lighting is weak, reds are rather hard to see by it. The spectra of different kinds of streetlight are shown in Figure 3.3.

These days, mercury streetlights are fairly uncommon in Britain, but they are still widespread in some parts of the US and other countries, though their use is declining and their sale is now banned in the US and EU. They are also used for decorative floodlighting, as with appropriate filters they can give a strong blue or green light.

If you want to look at the spectrum of a streetlight for yourself but lack the necessary equipment, you can do so using a compact disc. A CD is in effect a circular diffraction grating, even though its surface bears lines of tiny pits rather than grooves, and will produce a very satisfactory spectrum. Examine a streetlight by looking at its reflection in a CD (Figure 3.4). You will see that a spectrum is produced if you look across the disc, with the central hole between you and the light source. (Actually two spectra will appear, one on either side of the light source.) As long as the source is small in comparison with the width of the spectrum, you will see any spectral lines that are present.

◀ *Fig. 3.4 Line spectrum of a low-energy light bulb photographed by reflection off a CD. There are lines of different colors which combine to give white light. This is not a pure spectrum because the pits in a CD are not as regular as the lines on a proper diffraction grating.*

Fluorescent lighting

A variant on the mercury lamp is the familiar fluorescent or strip light, and indeed the low-energy bulb. This starts out life as a mercury-vapor lamp in the form of a long tube that gives out a considerable amount of ultraviolet light. The inside of the tube is coated with a material known as a phosphor that glows brightly when ultraviolet light strikes it. Different materials can be added to the phosphor to give different tones to the light, such as "warm white" or "cool." In everyday perception, these terms are the opposite of the temperatures that they represent – we associate red with hot and blue with cold.

The spectrum of a fluorescent strip light, particularly an older one, shows a strong green line and some other fainter lines, plus a pinkish glow which covers most of the spectrum. The full range of colors can therefore be seen in fluorescent light. The green line particularly affects the green sensitivity of cameras, which is why photos taken in fluorescent light can have a strong green cast to them, though digital cameras often compensate for this automatically.

Modern low-energy light bulbs are basically small versions of fluorescent tubes, though they tend to have different phosphors on them from the earlier tubes. Their spectrum has strong red and green lines, with the other colors being less dominant than in the older tubes. Computer and TV screens have a similar spectrum.

At one time it was the fashion to install fluorescent streetlights in city-center shopping precincts since their good color rendering made people's faces and clothing look natural. This is regarded as very important by lighting engineers – people hate lighting that gives their faces an unnatural cast. These days, however, high-pressure sodium and LED lighting has largely taken on this role. Strip lights are standard in large stores and offices and remain the most efficient means of lighting a large interior, though LED replacements are coming into use. Fluorescent lighting remains common in homes, particularly in kitchens, and this can have an effect on your observing as we'll see later in this chapter.

Low-pressure sodium lighting

Going back to my own personal streetlight, I was rather pleased when, sometime in the late 1960s, it was replaced with a low-pressure sodium lamp. This works in the same way as a mercury lamp, except that the gas used is sodium vapor rather than mercury vapor. Its spectrum consists of virtually just two lines of orange-yellow color, very close together. The wavelengths of these lines coincide quite well with the maximum daytime sensitivity of the human eye, so low-pressure sodium lights were regarded as being very efficient. They can give more light per watt of electricity than any other type, including the latest LED lamps. In the UK they became widely used, and are still common along minor and suburban roads. From the astronomer's point of view they are a good light source, since the twin spectral lines can be filtered out very easily, as described in Chapter 6. A low-pressure sodium streetlight is shown in Figure 3.5.

Unfortunately, nothing is ever perfect. There are drawbacks which prevent sodium from taking over from other forms of lighting. One is that it is quite impossible to see any color other than yellow by its light; all you can see are shades of yellow. Some people intensely dislike the color of sodium light for this reason, saying that it makes people's faces appear ghastly. It is hard to see why this should prevent it from being used on roads intended mostly for traffic. The reason often given is that it is important for the police to be able to distinguish the colors of cars, but in fact in low lighting conditions colors are hard to see anyway. And police cars, like others, are equipped with tungsten or LED headlights, which make car colors quite easy to see. Another technical drawback with the sodium-vapor lighting unit is that it is quite large. This becomes important when designing a lamp that shines only where you want it to. Finally, though their output is close to our daytime color sensitivity, at low light levels our eyes' color sensitivity is different, so they don't appear as bright as they should.

Paradoxically, the sodium lamp's very efficiency can give it a bigger impact on the night sky than mercury lighting. I remember hearing an American amateur visiting Britain comment that the orange glow in the

▶ Fig. 3.5 Close-up of a low-pressure sodium streetlight. The lamp itself is enclosed in a glass housing which is molded so as to refract the light sideways. This type of fitment has a considerable unwanted upward light spill.

night sky, caused by the widespread use of sodium lighting, was more visually distracting than the bluish color of mercury lighting. The eye is less sensitive to blue light than to yellow, so an orange-yellow night sky looks brighter than a blue night sky.

High-pressure sodium lighting

When an electric current is passed through sodium vapor at high pressure, the spectrum of the light emitted is broadened out from the two narrow orange-yellow lines into a band running from the red end of the visible region through to the blue end. Although the lights have a pinkish-yellow color, they give an acceptably lifelike color rendering. The lamps themselves are only two-thirds as efficient at turning electricity into light as low-pressure sodium lamps, but the color of the light is more closely matched to our night-time color sensitivity and the installation as a whole can be more cost effective because the lamps are brighter and cover a wider area. On an eight-lane highway, for example, fewer high-pressure lamps are needed to light the entire width than with low-pressure sodium.

To begin with, high-pressure sodium lighting was restricted to shopping areas – taking over the job previously done by fluorescent lighting – where a pleasing color was considered more important than cost. But in the 1980s it started to spread out from city centers, and can now be found on most major roads. Long stretches of Britain's highways are now lit by high-pressure sodium lights. As more of these lights have come into use, our skies are turning pink instead of orange.

Unlike low-pressure sodium, high-pressure sodium is virtually impossible to filter out completely. But from the astronomer's point of view there is one thing in its favor: because the light source itself is smaller than a low-pressure sodium lamp, it is easier to design a lamp housing that directs more of the light where it is needed. This has led to an important improvement in the fight against light pollution – the full-cutoff streetlight, about which more shortly.

Metal-halide lighting

By adding different gases to a mercury-vapor lamp, both the color and the efficiency of the lamp is improved. The result is known as a metal-halide lamp, and the white color and high intensity have made them very popular for such uses as illuminating sports facilities and other large areas. They can also be found in filling-station forecourts and some supermarkets, and occasionally as streetlights. Their light is white but the spectrum consists of numerous separate lines of different colors rather than being continuous. Some car headlights, known as xenon bulbs, are metal-halide.

LED lighting

We are all now familiar with LEDs (light-emitting diodes) as most flashlights now use them and most Christmas tree lights have these brilliant and yet low-energy bulbs. They are making their way into home lighting as well, and an increasing number of cars use LED headlights. Filling-station forecourts that might once have used fluorescent tubes or metal-halide lamps may now use LED lamps, and only if you look at them closely would you notice what they are. Further advantages are that LEDs are said to have a very long lifetime, reducing maintenance costs in replacing them, and also don't contain toxic mercury as do metal-halide and fluorescent lamps. And of course they are now increasingly used for streetlighting.

LEDs are highly directional, so there can be virtually no light spill from them. When a road in Harrow, Middlesex, was converted from high-pressure sodium to LED lighting, local amateur astronomers noticed the reduction in glare from that direction straight away. As I write these words, my local county council, Buckinghamshire, is replacing the streetlights along many of its routes with LED lamps. But it is too soon yet to say whether the night sky will turn from orange to white as a result.

Lighting types and night vision

Some lights are worse than others for astronomers. This is because the night vision of the human eye is controlled by a sequence of chemical reactions which alters its sensitivity to light. At night, your eyes are most sensitive to green and blue light, as emitted by mercury lights, so one glance at a bright blue light is enough to ruin your night vision for quite a while, until the chemical reactions have had time to restore it. Although the iris of your eye contracts and expands as well, this has only a very small effect on the overall sensitivity of the eye. The iris provides a short-term means of controlling the light received over a narrow range. Our eyes take time to become dark-adapted when we go out from a brightly lit room into the night. The iris expands almost immediately, but the chemical reactions that give us our most sensitive vision take much longer. You should allow at least half an hour to get dark-adapted; it is likely that your night vision will improve even after this.

The limited control provided by the iris will be appreciated by anyone who has their eyes tested by an ophthalmologist and is given eye-drops to dilate (expand) the pupils. When fully dilated the pupils may widen from about 3 mm in artificial light to almost 10 mm. Even when they are this wide, however, it is possible to go outside into bright sunshine with no more than a pair of sunglasses to make the light bearable. This experience shows that the iris has only a small effect on the eye's sensitivity: the chemical reactions are much more important.

It is because blue light destroys night vision that astronomers making notes of observations at the telescope habitually use red light. Red light has very little effect on your night vision, and you can carry on observing almost immediately after recording your observation.

You also need to take care of lighting before observing. Fluorescent lights and low-energy light bulbs can have a strong blue output compared with the old incandescent light bulbs. If you have them in any room before you go outside, or in the room adjacent to your outside door, you will have very poor night vision for some time. The solution is simply to switch off the light well in advance so that you do not have to look at it immediately before observing. (The rest of the household has to know that you do not want it switched on again, but this is a matter for domestic negotiation.) It may be worth putting a table lamp with a red or dim bulb in the room so that life can continue while you observe. By the same token, it is not a good idea to watch television or use a computer immediately before going outside. The phosphors in TV tubes and monitors can be just as damaging to your night vision as a fluorescent light. Computer software intended for use at the telescope, such as planetarium programs with telescope control abilities, often has a night vision option which gives a red screen.

Other people's fluorescent lights are another matter, and the sudden appearance of a kitchen or bathroom light at a crucial moment in your observing schedule can be very unwelcome, to say the least. About the only thing you can do is to anticipate the light being switched on and choose your observing site accordingly. Fluorescent lighting is also used in offices and stores, where it may be left on all night, and on filling-station forecourts. The light from these sources may spill upward to contribute to the skyglow.

Lighting a street

Streetlighting has to do different jobs in different locations. The level of lighting in a quiet suburban road, for example, does not have to be as high as on a fast major road. In city centers, lighting is as much for the benefit of pedestrians shopping in the evening as it is for drivers. In some areas it may be used to make pedestrians feel safe from attack by muggers.

The main aim in lighting a street is to enable drivers to see objects that might get in their way, such as pedestrians, dogs or other cars. Such objects do not need to be brilliantly lit, but enough light has to be put on to the road surface for them to stand out in silhouette.

▶ *Fig. 3.6 The same stretch of road with low-pressure sodium streetlights (above) and LED streetlights (below). The same lens and exposure details were used for both photographs. The road illumination is the same or brighter, but the light spill from the lamps is now minimal.*

Brighter lighting, often giving a good color rendering, is used on pedestrian crossings.

But there is more to lighting a road surface than putting a lamp on a post every so often. That would produce a pool of light below each lamp, but very little light in between the lamps. The farther you can make the light spread sideways, the fewer columns you need and the less it all costs. To do this, the glass housing of the light source may be fashioned so as to spread the light either side of the lamp, illuminating more of the road surface, as you can see in Figure 3.9 (b). In effect, it behaves like a lens. If you look at a road surface you may see a pattern of stripes caused by the focusing action of the housing. The lighting engineer has to make sure that the patterns from two adjacent lights overlap so as to produce as even an illumination of the road surface as possible.

Because such housings are designed to direct light sideways rather than downward, some of it can easily spill upward. (With the astronomer's favorite, low-pressure sodium lighting, the comparatively large lamp can lead to considerable upward spillage.) If the spread from each streetlight were not so great, more columns would be needed, costing more money and giving drivers more things to hit at the side of the road. So there is an economic benefit in making the light spread sideways. Since sideways lighting with low-pressure sodium lamps usually produces some upward spillage, it is easier to allow a little to

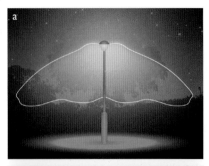

◀ *Fig. 3.7 This schematic illustration is based on polar diagrams produced by a streetlight manufacturer. Here, the extent of the white area in a particular direction from the lamp housing represents the intensity of the light as seen from that direction. Whereas the standard type of installation (a) sends some of its light upward, the full-cutoff design (b) directs all of its light toward the ground. (Based on information kindly supplied by Siemens Ltd.)*

shine upward than to cut it off, though with good design this can be avoided.

The manufacturers of the housings publish *polar diagrams* – plots of how the intensity of light varies around the whole fixture, which is known in the trade as the luminaire. To the surprise of astronomers, polar diagrams such as Figure 3.7 show only a small percentage of light directed upward. So what is the source of the problem? We know that our night skies glow with the color of streetlights, yet hardly any of their light is supposed to shine upward.

One reason is that the polar diagrams show the results of laboratory measurements on brand-new, clean luminaires. After only a short period of use out in the open, the glass of the housing becomes pitted and dirty, causing more of the light to be scattered upward than was intended. Another reason is that some light is reflected upward from the road and other surfaces. Yet another is that even light directed in the first place only slightly above the horizontal can eventually find its way up into the sky. So what starts off in theory as negligible ends up in the real world as a serious light-pollution problem. And it has also emerged that because the manufacturers have not been interested in upward-shining light, they did not bother to measure it.

What is the actual percentage that shines upward? According to figures obtained by Britain's Institution of Lighting Engineers, a typical value from a low-pressure sodium luminaire is 20–30% of the light that shines downward. A much smaller percentage is contributed by reflection from the road surface. A large proportion of British streetlights are of this type, though in other countries they may be less common.

Full-cutoff lighting

Largely as a result of pressure from astronomers, the lighting industry became aware that the upward light spill was not only causing light pollution but reduced the efficiency of the luminaire. Collectively, poor design resulted in whole power stations being needed simply because the light wasn't going where it was needed. Originally, high-pressure sodium housings were designed in the same way as those for low-pressure sodium, but new designs made it possible to change the way in which the light is spread sideways. The beam of light is shaped by reflectors inside the housing rather than the lensing effect of the glass. There is no possibility of light escaping upward, and the beam cuts off sharply at an angle. These are called full-cutoff or flat-glass streetlights, and the lamp itself is set inside a housing with a flat, horizontal glass window. A streetlight of this type is shown in Figure 3.8. LED lighting by its nature uses tiny individual light sources, and this too is easily made to be full-cutoff.

The benefits to both astronomers and drivers are considerable. From the same level as the lamps you can see that the beams shine downward, and that the only light reaching the sky is reflected from the road surface. From the driver's point of view, full-cutoff lighting is free of much of the glare that comes with older types. This really is a boon for driving in the rain, since much less light is scattered by droplets or dust on the windscreen, and the driver has a better view of the road ahead.

Another group of people are happy, too. Those who live close to main roads in rural or semirural areas often object to the illumination of their houses and gardens by the street-lighting. With the new lamps, the spill-age is much less than before. At one time these lights cost more to install, because they needed to be mounted higher up in order to give the same spread of light as before. But modern designs have made these lamps as energy efficient as the older types, even when used in quiet residential roads.

▶ Fig. 3.8 A flat-glass, low-pressure sodium streetlight – the astronomer's preferred design. The design of this unit prevents any upward light spill. The photocell that switches the light on and off can just be seen on the top of the unit.

◀ *Fig. 3.9 (a) A street illuminated by standard low-pressure sodium lights, with poor control of light spill. On a misty evening the upward glare from each lamp is clearly visible. Very close to where this picture was taken, the street crosses a motorway lit by full-cutoff streetlights (b), photographed under the same misty conditions as in (a). From this vantage point the lamps themselves are almost invisible, and the beam pattern shows very clearly that there is no upward light spill.*

Full-cutoff streetlighting is now installed on many highways throughout the world (see Figure 3.9), and as lighting engineers have become more aware of its advantages it has spread to other roads as well. In Britain, full-cutoff luminaires are obligatory on major roads where new lights are being fitted.

For a while, it seemed that the increasing use of full-cutoff lights on major roads would reduce light pollution. But every amateur astronomer who has been observing for a number of years knows that the light pollution has got steadily worse, not better. Why?

The very efficiency of full-cutoff fittings has meant that more light now goes on to the road rather than into the sky. Although the reflectivity of the road is usually less than 10%, this still has a part to play, and some roads are now illuminated very brightly. More roads are now illuminated, usually on safety grounds. For example, some stretches of the London orbital motorway, the M25, were until 2012 not illuminated and had six lanes (three in each direction). Now those stretches have eight lanes, and are brightly illuminated. But it's not just roads – there are other sources of light pollution as well.

Other forms of light pollution

Streetlights are simply the most visible form of light pollution. There are many others, of varying degrees of severity. Probably the most significant are floodlights used to illuminate sports grounds, buildings, and parking lots; lighting at shopping precincts, industrial plants and filling stations; and advertising signs and security lighting. Ordinary domestic lighting usually plays only a very minor part. Less common but just as bothersome if nearby are such one-offs as laser shows. The relative contributions of these various sources to the total sky brightness varies very much from place to place.

Floodlights at sports grounds now largely use metal-halide lamps. There is a regrettable tendency to point the lamps sideways from the edge of the area (Figure 3.10), with the result that they shine as much light upward as downward on to the pitch. For this reason they are a major source of light pollution, as one upward-pointing light can create as much skyglow as hundreds or maybe even thousands of well-directed streetlights. As a result, they can be seen over a huge distance. An increasingly common location for floodlighting in Britain is the golf driving range, which can be in operation for many hours in the evening.

Many buildings are now floodlit at night, and not just major land-marks and public buildings but also churches and office blocks. Such floodlighting is partly for prestige and partly for security, because an illuminated building is regarded as being more difficult to approach at night without being seen. The cheapest way of arranging the flood-lights is to put them on the ground pointing upward – very bad news for astronomers (Figure 3.11).

Some lighting regulations do specify that 90% of the upward-directed light should hit the object being illuminated, but even this means that a vast amount of powerful lighting goes into the sky. The use of LED lights could improve on this figure, as the sources are much more directional. But unless the controls are very strict, and the installations well designed, it will remain easy just to plant a few floodlights around a building and blaze away. Another pernicious source of light pollution is the uplighter used for decorative effect, such as on trees. Even the National Trust, an organization devoted to preserving the UK's heritage and a great advocate of conservation,

▶ *Fig. 3.10 How **not** to light a sports field. This installation sends as much light into the sky as it does on to the pitch and can be seen for miles.*

◀ *Fig. 3.11 Floodlights such as these on a golf club building contribute more light pollution than a whole estate of roads and houses. The lighting on this single installation shows up clearly on photos taken from the International Space Station from orbit.*

▶ *Fig. 3.12 A globe light in a parking area, a good example of an extremely poor lighting unit. The light source is mercury, which is inefficient compared with sodium, and more light goes upward than downward. Most businesses on this site have had to install separate lights to illuminate their parking lots.*

uses uplighters on many of its country properties, often where there isn't even a tree to shine on!

Parking lots, particularly around supermarkets, and pedestrianized shopping areas are often lit by globe lights – transparent or opaque globes with a lamp inside. As Figure 3.12 shows, they are singularly useless at doing the job of lighting the ground since their light is totally undirected. They will illuminate any nearby buildings, which streetlighting engineers sometimes consider a good thing since it is supposed to enhance the appearance of the area. But a globe light is one of the most inefficient means of lighting there is. They look pretty by day, and in the architect's drawing; but by night, glaring over an empty parking lot, they are a disaster. Some designs have the dubious distinction of throwing more light into the sky than on to the ground, since the base of the lamp fitting can cast a large circular shadow. This deep pool is a good place for muggers to hang out, unseen against the light from the lamp itself, but with a good view of their prey.

Because globe lights are so inefficient, they are commonly used in clusters with several units on one column. But some types of globe light are designed to throw the light downward, using internal reflectors. Not only are these better from the astronomer's point of view, but they also illuminate the ground much more effectively.

These days, security lights are common. Quite often they are mounted on the wall of a building, and arranged so as to illuminate as large an area as possible. This usually means pointing them outward so as to deter people from approaching at night. Security lights for small businesses, homes and parking lots often have 500-watt tungsten or tungsten-halogen lamps. Actually, these are very inefficient and expensive to run, and quite often are used in an ill-planned way. Their very brightness usually means that there are also pools of inky darkness, and the fact that they are on the building pointing outward means that anyone looking toward the building who might be in a position to spot an infiltrator will instead be dazzled by the glare of the lamps.

By and large, the effects of security lights are quite localized, frustrating the individual observer rather than adding greatly to the sky brightness, since their contributions are small compared with that of floodlights in urban areas. A bright light shining in your eyes, even from miles away, is just as disconcerting as a bright sky.

David Crawford, of the International Dark-Sky Association, assessed the main contributors to light pollution for the first edition of this book in the 1990s, and there is no reason to suppose that it has changed much:

"We have estimated that about one-third of the skyglow problem comes from streetlights in the average community. Another third comes perhaps from sports lighting (though this can be really bad in some locations), advertising lighting, and such. The other third covers all the rest, including private lighting. It varies with the size of the community and other things. Here in Tucson, all street and parking lot lighting is full-cutoff, and so the 'third' must be less. We estimate that bad sports lighting and bad advertising lighting are now the dominant problems. The industrial yards are a real problem in some places, as the lighting is almost always poor. It is not too bad here in Tucson, though. Some locations, like Oakland, California, have very bad lighting at docks too."

What can be done about light pollution?

At one time, it seemed that astronomers were powerless to do anything about the encroachment of all the various sorts of lighting. Apart from a few areas near major observatories, such as in La Palma in the Canaries and Tucson, Arizona, astronomers had a very weak voice indeed. After all, they were mostly either amateurs, or professionals who could presumably visit telescopes on remote, dark sites. They were very much in the minority compared with the large number of 'users' of streetlights: in theory all drivers, all pedestrians and anyone using premises lit at night – in short, most of the population, whether they were aware of it or not.

But although the problem seems to get worse by the year, there are signs that the tide is turning. In Britain and America, substantial improvements have been made by such organizations as the Campaign for Dark Skies and the International Dark-Sky Association. The aim of these groups is to raise awareness of the problem at national level, and they have been amazingly successful in doing so, considering the very limited resources available to them. Today, all lighting manufacturers are aware of the problem of light pollution, and refer to it in their publicity. But despite this, there are economic arguments which weigh heavily against the astronomer.

The ways in which streetlighting can be controlled break down into control of the type and direction of the lighting, control of the time for which it is on, and control over the need for it in the first place. The circumstances governing each case vary from location to location. Full-cutoff luminaires, of whatever type of light, are preferred as they shine very little light upward. As most streetlighting installations have a lifetime of around 30 years, it would be uneconomic to replace existing fittings with full-cutoff ones. So astronomers cannot hope for an instant solution to the problem of streetlight glare.

Dark skies after midnight?

Every cloud has a silver lining, and the recession in the mid-2000s made many councils rethink their lighting policy and decide that running streetlights after midnight when hardly anybody was around to use them was a waste of money and electricity. There was the inevitable reaction from people who suddenly found that they could no longer find the keyhole of their front door on the odd occasions when they returned after midnight, or who worried about the effect on their safety, but by and large the switch-off was pushed through and even some quite major roads are now unlit after midnight with no ill effects. This can have a noticeable effect on sky brightness. Martin Lewis saw this when he bought an all-sky camera and made time-lapse videos of the sky from the suburbs of St Albans, a city of about 60,000 people. After midnight the drop in sky brightness was obvious (Figure 3.13).

Of course, it only takes a slight swing in finances, or a change in the political color of a council keen to curry favor with its voters, and the lights will go back on again.

Many non-astronomers, if asked, would say they prefer the streetlights to be on all night if it means safer streets. It is often argued that if one life is saved each year then the expense is well worth it. So by keeping the lights shining, civic authorities are apparently keeping most of their customers happy, which means a great deal to them. Some take a rather mercenary approach to the cost-benefit analysis of streetlighting, for example setting the cost to the community of keeping streetlights on against the hospital bills for accidents supposedly caused by turning them off. Authorities are concerned about their possible financial liability should a citizen be able to persuade a sympathetic judge that an accident or break-in was the result of a streetlight being turned off.

Streetlights and crime

While streetlights on main roads are intended to make driving safer, those in residential areas have an additional purpose – to make the streets feel safer to pedestrians. Local news bulletins and newspapers are dominated by stories of increasing crime, from rapes and muggings to burglaries and car theft. Even in areas where the worst crime for years has been someone riding a bike without a rear light, many people – particularly the elderly – are afraid to go out at night.

Streetlights are seen by the average householder as a defense against this increasing lawlessness, and it would take a powerful argument indeed to persuade them otherwise. The facts are that more residential break-ins happen during the day, while more crimes against the person are committed in well-lit areas where victims are easy to find. In the UK, a study by the Home Office confirmed that intending criminals pay little attention to the presence of streetlights, but also found that people feel safer with streetlights on, though in fact they have little effect in reducing the crime rate. Actually, a study of two towns in

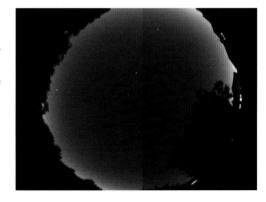

▶ Fig. 3.13 Before and after the switch-off. A comparison of all-sky photos made before (left) and after midnight in St Albans, Hertfordshire, UK, shows the striking difference made by switching off many streetlights. There is 7 minutes' difference between the two shots.

Essex found that when streetlights were turned off at midnight, night-time crime actually dropped considerably. But it's no use astronomers pointing out that lack of streetlights does not cause crime – in any debate, they will lose hands down to the majority view.

Positive action

All this makes it sound as if the best thing the urban astronomer can do is to head for the country or take up another hobby. But there are ways of lighting our streets and buildings to make them safer without adding to light pollution.

Given that many people feel safer if there is light around, despite plenty of evidence to the contrary (see below) there is no point in you simply demanding that streetlights be switched off or taken down – you will be dismissed as being in a tiny minority. It is much more effective to ask that lighting should be used efficiently, so that not only are the streets kept safe, but the cost is kept to a minimum. And to a certain extent the vocal demands for more light are tempered by some public sympathy for the notion of dark skies, particularly in country areas. A non-astronomer friend of mine was on a council residents' advisory committee when the topic of streetlighting came up. One person said that we needed more lighting, but when my friend voiced his opinion that as a country resident he preferred to see dark skies, he found that the majority of the committee agreed with him. And some places actually prefer not to have streetlights (Figure 3.14). All things being equal, many authorities will be quite happy to install astronomer-friendly lighting. Today, full-cutoff lighting is widely accepted as being not only good for the skies but good for the public purse as well.

National publicity is raising the public awareness of light pollution, but only local action can get things done at neighborhood level. As this chapter has shown, however, it is important to have facts and figures available which prove your point, otherwise local authorities will simply dismiss your demands. Join the International Dark-Sky Association or the Campaign for Dark Skies, and ask for their help first.

Every council and local government authority has an official, often a professional lighting engineer, who is responsible for the public light-ing in the area and who sees all planning applications for new schemes. This is the person to whom you should put your point of view. Ask for full-cutoff lighting to be installed wherever it is practical, and ask also that consideration be given to switching off non-essential lights after midnight. Your appeal may not have an immediate effect, but it will have raised the issue. Put it this way: if there were no demand for darker skies, no one would do anything about it.

Your local lighting engineer, however, does not have control over all the lighting in the area. Lights on private property are often installed with little professional skill. Once they are up, there is a strong economic argument against removing them. So, if possible, keep an eye on all planning applications which may involve lighting – in the UK they are usually published in local papers and online – and make your views known in advance. If a new supermarket or shopping area is being planned, ask for details of the proposed lighting.

When it was learned that a new road with its associated lighting was planned to run near the observatory of the Croydon Astronomical Society in south London, it seemed that all the society's work might be wasted. Nevertheless, a case was submitted which emphasized the educational use of the observatory, requesting that dark-sky standards be applied to the lighting – which in the UK means full-cutoff high-pressure sodium. To the society's delight, their case was accepted without question. They were told that this was because they had made their views known in good time, before detailed plans were made, so they could be taken into account.

In the case of security lighting, you could ask for it to be controlled by infrared beams so that it comes on only when there is an intruder, rather than being left on permanently. This is a contentious area. People who feel they need security lighting do not want to be told by others how to control it. They could argue that the very presence of bright lights puts people off (though there is some evidence to the contrary). Only if the lights are incandescent or LED can they be switched on instantly – gas-discharge lamps such as sodium or mercury take some time to reach full brilliance.

Miss no opportunity to put your views across. Write to your local newspapers pointing out that it is up to all of us to control lighting so that the skies remain dark while the light that is needed goes where it should. Write to your MP or Congressional representative about it. Write to the managers of local stores with bad lighting, asking that in the cause of good customer relations and a green

▶ Fig. 3.14 The residents of Theydon Bois, about 24 km (15 miles) from central London, have repeatedly voted not to have streetlights, fearing that these would destroy the ambience of the area.

image they could gain useful publicity by improving the quality of their fittings. The Campaign for Dark Skies gives Good Lighting Awards to businesses that change their lighting for the better. When Gloucester amateur astronomer John Fletcher requested that a local pub remove their high-intensity upward-pointing searchlights, they agreed and received both the award and good publicity.

As well as corporate bad lighting, your immediate neighbors may have interior or exterior lights that ruin your observations. Sometimes the perpetrators of bad lighting are totally unaware of the nuisance they are causing. A tactful approach pointing out the problem can sometimes work. But often, people see any such comments as unwelcome criticism and interference with their life and safety, so you may need more diplomacy than a whole embassy of ambassadors could muster. It might be a good idea to invite them round to look through your telescope, if you have one, to see a few of the wonders of the skies and to discuss ways in which controlling their lighting could help you see more.

Light pollution and the law

There are now laws governing the use of outside lighting in the UK and some other countries. The situation in the US is less clear because each state has its own laws rather than a national policy.

In the UK "light nuisance" is now subject to the same criminal law as noise and smells. It applies to "artificial light emitted from premises so as to be prejudicial to health or a nuisance." The main value of it is to prevent strong lighting from flooding into people's bedrooms, and from shining into drivers' eyes, but the unwanted intrusion of a neighbor's light which interferes with your own lifestyle could also be considered a nuisance. But taking your neighbor to court is a last resort, and the usual advice, if you can't sort things out with a chat, is to keep a log of occasions when their lighting has been a nuisance and then involve the local environmental health officer. Unfortunately, some industrial premises such as railroad yards and ports are excluded from the legislation, and unless the lights are shining on to your property rather than just creating unwanted skyglow in the area there may be nothing you can do about it. There is plenty of information about this on the Campaign for Dark Skies website, www.britastro.org/dark-skies.

But when all is said and done, your skies will probably still be bright. You will have to come to terms with the sad truth that light pollution will be around for a long time to come. So the next thing the urban astronomer must do is to choose the right weapons and targets to make the most of what is still available.

4 · CHOOSE YOUR TARGETS

Urban astronomers must be more selective than their country cousins in what they observe. Generally speaking, the brighter objects can be seen just as well from the city as from the country; it is the fainter objects that cause the problems. The determined urban astronomer nevertheless has much to choose from, including many of the sky's showpieces. In this chapter I describe what you can observe from under light-polluted skies, in each case giving a brief guide to getting the most out of your observations – if you want to. There is no reason to feel that you have to look at everything there is to see. To be honest, the vast majority of amateur astronomers rarely observe seriously, but simply get a kick out of viewing their favorite objects. I'll run through the celestial targets in a somewhat arbitrary order of brightness and interest, starting with the Sun and the Solar System.

In virtually each case, technology in the form of advanced imaging cameras and computers has transformed the ways you can observe. But most people want to see the objects before they start to go down the technological route, so I've explained how much you can see with your eyes alone and using telescopes or binoculars before going on to the more advanced techniques.

The Sun

As an object for study, the Sun has a lot going for it. You lose no sleep observing it, you can usually keep warm, it is there in the sky more often than many other astronomical bodies, you do not need a large telescope, and its features change daily, if not hourly.

Being in an urban area is no real drawback to the solar observer. You're as likely to have a building in the way of your target as the country observer is to have a tree, and the seeing can be bad in either location. While it is true that the best seeing demands mountain-top or lakeside sites, the average city location is probably not much worse than the average country one. The best seeing is usually in the morning, as early as you can get to it. Bad industrial pollution reduces the contrast of the Sun's image, but not normally by so much that you cannot observe. All in all, however, the Sun is an ideal target for the urban astronomer to study.

The Sun is so bright that a large telescope is a liability rather than an asset. A telescope as small as 50 mm (2 inches) in aperture is more than adequate, and 75 mm (3 inches) is often said to be ideal. The problem is cutting the light down rather than capturing enough of it, which is the difficulty with most other celestial objects. The Sun is unique in presenting dangers to the observer. Looking at it directly through a telescope or binoculars is like using a burning glass on your eye, with temporary or permanent blindness the result.

Even looking at the Sun with the naked eye is dangerous, and our brains usually prevent us from doing so when it is too bright. Only when it is not painful to look at – such as when it is very low in the sky with its light significantly dimmed by haze – is it safe to observe without a filter, and even then there may be people who are particularly at risk, so do not take this as medical advice. There are records of the visibility of sunspots in such conditions, notably from the Orient, from long before Galileo first drew the attention of Renaissance Europe to them in 1610. Today you may be tempted to use a dark filter, but you should steer clear of anything which is not specifically designed for solar work. There is no guarantee that a filter that cuts down visible light will also cut down the infrared, which will burn your retina in the absence of the involuntary reaction against a bright light that would otherwise close the eye.

I must now deliver the obligatory warning against using the solar filters that were once provided with many small telescopes. These are made of dark glass and generally screw on to the eyepiece. There is a grave danger that they will shatter unexpectedly, especially if they are used with the full aperture of the telescope. Usually supplied too is a cap that fits over the top of the tube and cuts down the aperture to about 50 mm (2 inches) for observing through the solar filter. You should **never** observe the Sun with an eyepiece filter alone. It is positioned at the greatest concentration of the Sun's rays, so it inevitably heats up. Even a hairline crack, which you may not notice, can let through dangerous amounts of heat. This applies to naked-eye use as well as telescopic work – although a filter is unlikely to crack when simply held up to the Sun, it may be cracked already.

◀ Fig. 4.1 When projecting the Sun, stop the aperture down to around 50 mm to avoid excess heating of the eyepiece. With a Newtonian reflector, as here, cut the hole in the stop so that the Sun's light misses the secondary mirror support. But spot the mistake here – the image from the uncapped finder telescope is in danger of burning someone's sleeve.

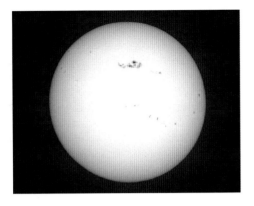

▶ *Fig. 4.2 The Sun photographed in white light, showing several sunspots. This view was obtained by photographing the projected image in a darkened room, using an ordinary camera alongside the telescope.*

Fortunately, these filters are much less common than they used to be, but a quick search of the Internet showed several second-hand telescopes offering them and various Chinese manufacturers providing them on new products.

Just pointing your telescope at the Sun can cause problems. Always keep the lens cap on the finder telescope, if you have one, to prevent accidents. Bear in mind that the beam from the eyepiece of either finder or main telescope can burn. The best way to align your telescope on the Sun is to watch the shadow it casts and move it until the shadow is as small as possible.

The cheapest way to observe the Sun is to project its image on to a screen, as in Figure 4.1. You can do this with any kind of telescope. Regular observers use boxes which keep stray light off the screen, giving much better contrast. It used to be said that projection is also the safest way to observe the Sun, but there are still risks. Someone may inadvertently look into the eyepiece, or, more likely, you can damage the inside of the eyepiece. Inside each eyepiece is what is called the field stop, a circular aperture that defines the edge of the field of view. Very often this is made of plastic, even if the eyepiece barrel is metal, so it can easily melt if subjected to the intense solar image. You really need an old eyepiece that is not only expendable but also has a metal field stop. But if you can overcome these problems, you will be rewarded with fine views of our nearest star. Sunspots stand out with good contrast, and the bright areas known as faculae also show up well (see Figure 4.2).

Solar projection allows you to make drawings of the Sun's features, which is the usual means of recording their positions. Not only can you detect the Sun's rotation, but you can also see how the spots change from day to day. For drawing the Sun, prepare sheets with a standard blank disk with a diameter of 152 mm (6 inches). It is possible to mark the positions of features on a blank disk by simply projecting the Sun's image on to it and marking them in, but a better method is to project the image on to a faint pre-prepared grid. You then transfer the locations of the features on to a blank disk under which you have put an identical grid ruled with bold lines that show through the paper.

These days, a more popular method of observing the Sun is to use a dense filter over the objective of a refractor, or over a hole cut in a mask fixed securely at the top end of a reflector. The cheapest of these filters are made from thin aluminized monomer film, such as Baader AstroSolar™, and are definitely not suitable for use at the eyepiece end. You can buy this in sheets and cut it to your requirements. It is available in two densities, visual and photographic, the latter giving an image 100 times brighter, which is unsuitable for direct visual observing and is only meant for high-magnification imaging.

The film is delicate, so before using such a filter check that there are no tiny pinholes in the coating that could let through light. Most importantly, make absolutely sure that the filter cannot be dislodged under any circumstances. Your telescope may already have an aperture in its tube cap that restricts the aperture to about 40 mm, which is designed for solar observing. Cover the inside of this aperture with solar film and you have your filter (Figure 4.3). Don't worry if the film isn't completely flat. It's so thin that crinkles have no effect on the image at all (though bad creasing which might cause pinholes is another matter). With this, you can view the Sun directly, but drawing what you see is then a matter of estimating the position of the spots, which can be tricky.

Using such a solar filter, particularly a home-made one, is rather like driving at speed in the fast lane – it is perfectly safe most of the time and you can feel quite secure, but if the unexpected happens the results can be sudden and disastrous.

◀ Fig. 4.3 I taped a small piece of Baader AstroSolar™ to the inside of the aperture in the dust cap of this 80 mm refractor to turn it into an excellent and safe solar telescope. The image it gives is pure white – the actual color of the Sun.

▶ *Fig. 4.4 A glass solar filter fitted to a Meade ETX Maksutov telescope gives an excellent solar image which you can view in perfect safety through the telescope.*

When there are other people around there's always the risk that someone might helpfully remove what they think is the lens cover, for example.

More expensive filters also use reflective coatings, but on optical-quality glass mounted in a cell that can be firmly fixed to the objective (Figure 4.4), which is much safer as it avoids any risk of the filter dropping off. But these can also be damaged by rough handling. Always follow the manufacturer's instructions to the letter when using solar filters.

Don't try to find cheap ways of filtering the Sun's light. Food packaging made from aluminized Mylar should not be used to make a solar filter: it was intended to protect the food, not your eyes. I have seen pieces of colored plastic offered for sale as solar filters with extravagant claims that could not be backed up. When one of these filters was put in front of a TV remote control, which sends signals at infrared wavelengths, the TV still responded to the commands, showing that the filter transmitted at least some infrared. This is by no means a definitive way of selecting materials opaque to infrared – that can be done only by detailed laboratory testing. It did reveal, however, that there are no established guidelines for safety when observing the Sun. Welder's glass must be regarded with suspicion since it is designed for a different purpose. However, for looking at the Sun without a telescope to check for naked-eye sunspots, or when viewing an eclipse, you can use a shade 14 welder's glass, though this may turn the Sun a rather unappealing green!

It is not necessary to draw the Sun each time you observe it. A simpler way is merely to count the number of active regions and sunspots. An *active region* is an area in which there is a single large spot, or several spots all within 10° of one another in terms of solar latitude and longitude. To begin with, you need to record the positions of all the spots visible and then work out how many are within 10° of each other. With practice you can do this simply by estimation.

The daily tally of active regions and sunspots may be used to work out the sunspot number, Wolf number or Zurich number. Simply multiply the number of active regions by 10, and add the number of sunspots – spots belonging to an active region as well as lone spots. So if there are two active regions and seven individual sunspots in all, the sunspot number for the day is 27. Estimates vary from person to person, and with the equipment used (close spots may be difficult to distinguish) so you may have to apply a factor to your own estimates to bring them into line with everyone else's. You can compare your results with those published daily on www.spaceweather.com, a good source of information on all things solar.

The way the sunspot number changes over the months and years is of more than academic interest. The number varies, with a very obvious cycle of roughly 11 years. At times there are few, if any, spots visible on the Sun, and those that can be seen are all at high solar latitudes. A few years later the Sun will be covered in spots, mostly closer to the equator. At the peak of the solar cycle the spots are at their most numerous and other solar phenomena also reach a maximum. At these times, frequent bursts of radiation from the Sun can have a significant effect on Earth. Aurorae (the northern and southern lights) are much more common, and occasionally long-distance radio communications and even power supplies are disrupted. There may be extra drag on satellites in low orbits, causing them to spiral down to a premature end. Astronauts in space may be subjected to additional radiation hazards.

The Sun at other wavelengths

Straightforward white-light observation of the Sun's disk, as described so far, is the bread and butter of the solar observer. But enthusiasts have long made use of the ample supplies of radiation from the Sun across the whole electromagnetic spectrum to single out other wavelengths for study.

The Sun's visible spectrum, like that of most other stars, consists of a bright rainbow of colors crossed by dark lines produced by atoms and molecules in the Sun's atmosphere. These lines are named *Fraunhofer lines* after the German physicist Josef von Fraunhofer, who was the first to map and study them at the beginning of the 19th century. Hydrogen is by far the most abundant element in the Sun, and the three strong hydrogen lines in the visible part of the spectrum are favorite targets for solar observers. The lines are the first in a series of lines called the Balmer series, and are assigned letters of the Greek alphabet. *Hydrogen-alpha* (usually called H-alpha) has a wavelength of 656.3 nm, corresponding to a deep red color near the limit of the human eye's red sensitivity. It can therefore be hard to spot in the Sun's spectrum, even though it is a very strong line.

Although the Sun's atmosphere absorbs light at this wavelength, there is still plenty that gets through. The giant protrusions from the Sun's atmosphere known as prominences actually shine with this color. They come into view during total eclipses of the Sun, looking like pink flames around the limb (the edge of the disk). The pink color is a combination of the deep red hydrogen-alpha, the less strong green hydrogen-beta, and the weaker blue hydrogen-gamma line.

These days you can buy filters which transmit hydrogen-alpha only. The cost is not trivial: for the filter alone, expect to pay about the same as for a well-equipped 150 mm (6-inch) reflector for even a basic one. There are much more expensive versions that transmit a narrower wavelength band, say 0.7 nm, costing more than a top-end professional camera. Do not mistake these *narrowband filters* for those intended for deep-sky or CCD photography – also referred to as H-alpha filters – which simply cut out all radiation other than the deep red. Note, too, that the exact wavelength transmitted by some filters depends on their temperature, and may also vary with the filter's age. Beware of second-hand filters that may need to be heated to extreme temperatures before they transmit H-alpha!

Rather than buy the filter set, many people opt for a ready-made solar telescope. Instruments such as the Lunt LS35 and the Coronado PST are ready-to-go solar telescopes, tube only, each costing as much as a 300 mm (12-inch) Dobsonian telescope but for an aperture of 35 or 40 mm respectively. Nevertheless, these solar telescopes are very popular.

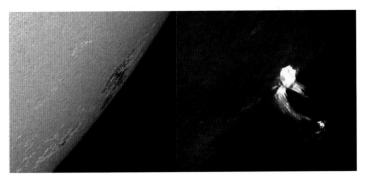

▲ Fig. 4.5 A comparison of the same part of the Sun in white light and hydrogen-alpha. The prominence is not visible in white light, but the sunspot is hardly visible in H-alpha. The bright area is a flare, which remains close to the solar surface, but the prominence extends up into the solar atmosphere.

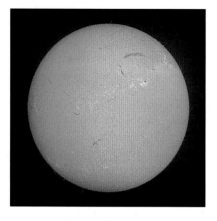

◀ *Fig. 4.6 A city location is no hindrance to solar observing. This whole-disk picture was taken from Edgware, London, on March 23, 2011, through a double-stacked Lunt LS60T using a DMK 41AF02.AS webcam.*

With an H-alpha filter or telescope you can see prominences around the limb of the Sun, if any are present. Most of the time you will see only a few rather insignificant spikes sticking up, or maybe nothing at all. They may change shape in a matter of minutes, and may detach themselves altogether from the limb. Do not expect to see anything like those movie sequences of giant flames leaping from the Sun. They were taken at times of extreme solar activity, and have been greatly speeded up. The most spectacular prominences are usually associated with large sunspots (Figure 4.5).

The disk of the Sun as seen in H-alpha has a much more mottled appearance than in white light (see Figure 4.6). The sunspots are not so easy to see, but their locations are usually obvious from the more disturbed nature of the area. As well as H-alpha filters you can get calcium filters, which isolate the blue calcium line.

The Sun can also be observed at radio wavelengths, a topic covered in Chapter 8.

The Moon

People expect a lot from their first view through an astronomical telescope. They want to see the dramatic swirls of gas clouds in space, the vivid festoons of color on Jupiter, the fantastic landscapes of alien moons. In most cases, the reality is disappointing by comparison – except for just two objects. One is the Moon, especially at first quarter. (The other is Saturn – see page 66.)

No matter what instrument you use, from binoculars to a high-power telescope, the Moon can be a spectacular sight. It is a challenge to try to observe the Moon when it is a thin crescent, which demands a good horizon and clear skies at just the right time. Just after sunset light pollution does not interfere, though urban pollution does. An observer in a high building stands a better chance than one at ground level, so a suitably located city observer could have more success than one in the country. Binoculars are helpful in picking out a very young Moon. The real test is to see a Moon which is less than 24 hours old.

Telescopically, the Moon provides hours of enjoyment. I like to gaze along the limb, where I can see the true slopes of the mountains, as if I were on the surface itself. The harsh lighting and long shadows of the terminator – the shadow line – give the impression that the Moon's landscape is more rugged than it really is (see Figure 4.7). At the limb, you can see that the slopes are actually gentle and rolling, rather than jagged.

The Moon is undeniably a crowd-puller, but what can you actually do with it? As recently as the middle of the 20th century, there were still parts of its visible face waiting to be mapped by non-professionals. Amateurs were still making discoveries, and hotly debating the origin of the Moon's features. Now virtually the entire Moon has been mapped, men have walked its gray rolling plains, and scientists have analyzed its rocks in Earth laboratories. Is there anything left for the amateur to contribute?

Surface changes?

There is one very good reason for watching the Moon. Occasionally – very occasionally – odd things seem to happen, for there have been reports of unusual glows or obscurations. These are grouped under the general heading *transient lunar phenomena*, or sometimes lunar transient phenomena (TLPs or LTPs). Like reports of UFOs, the Yeti or Bigfoot, they can seem reliable or even convincing, but are tantalizingly hard to explain. Are they escapes of gas from the interior, or the result of meteorite strikes, or simply illusions? We still

▶ *Fig. 4.7 With almost any form of optical aid the Moon is a spectacular sight, particularly when close to a quarter phase when the terminator region is thrown into sharp relief by the low angle of illumination from the Sun.*

◄ *Fig. 4.8 The south polar regions of the Moon viewed through my 200 mm Meade LX90 telescope. You can see that the slopes of the craters are actually fairly gentle.*

await some more undeniable events, seen independently by several observers.

There are certain craters, Aristarchus, Grimaldi and Plato among them, whose names keep cropping up in TLP reports. The Aristarchus area in particular, including the nearby Herodotus and the "Cobra Head" of Schroter's Valley, has given rise to a third of all TLP sightings (Figure 4.9). This is a fascinating area of the Moon – Aristarchus is the brightest lunar crater, and one of the most recently formed. In 1963 reddish glows were reliably reported near Aristarchus, and a brightening was photographed during a solar flare. Spectrograms have been recorded of other glows, though these have been disputed.

There is no doubt about one form of transient event – meteoroid strikes on the Moon. The Moon has no significant atmosphere, so there can be no meteors as we see them in the Earth's atmosphere. These are actual impacts of bodies with masses of the order of tens of grams or more – from gravel up to quite large stones. Such bodies would burn up in the Earth's atmosphere if they were to encounter it, but on the Moon their space velocity is converted instantly into light and heat when they strike. Not only have these events been seen visually, they have also been recorded on video so they definitely happen.

However, using your camcorder or webcam in the hope of videoing a flash is not likely to be successful. The recommended setup uses a large telescope and a low-light video camera, with accurate timing. Do a web search for NASA lunar impacts to find out more.

Should you ever see an unusual and continuing phenomenon, you should alert other observers if possible, but avoid giving them clues as to what is happening, or exactly where, to prevent bias in their reports. Observations of color are best substantiated using a reflector or a top-quality, perfectly color-corrected refractor so as to avoid any suspicion that the color is an artifact of the instrument. A filter of contrasting color can help to enhance the contrast of any colors seen – a blue filter will make a red patch appear unusually dark, for example. A pair of red and blue filters, used in succession, can make a colored area appear to blink in brightness. The British Astronomical Association's Lunar Section and the Association of Lunar and Planetary Observers collate observations of TLPs.

Although it is worth keeping an open mind about such phenomena, it has to be admitted that very few people have ever seen a TLP, and that to all intents and purposes the Moon's face is unchanging. But there is another good reason for observing it: it is ideal for honing your observing skills. In these days of high-tech imaging it may seem quaint to go out and draw part of the Moon, but it is an excellent way of training your eye.

Set yourself a task – mapping a certain area, for instance. Make a detailed drawing every so often, and you will find that your perception of the different features changes. What seemed at first to be a hillock turns out later, when the lighting is different, to be a craterlet. The Moon rarely presents exactly the same face to us each month, but wobbles so that we see slightly more of one side one month, then slightly more of the other side the next. This effect is called *libration*. The Sun's angle above your chosen area will also vary from, say, one first quarter to the next, so you can get a huge range of aspects of just one small part of the Moon over a period of time.

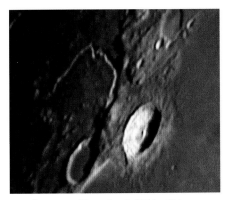

▲ *Fig. 4.9 When Peter Grego drew the craters Herodotus and Aristarchus on May 31, 1985, using a 175 mm refractor he saw a hill on the floor of Herodotus (left-hand crater), although it actually has a flat floor, as shown on my photo (right) through my 200 mm* *telescope on November 1, 2006, which has nearly the same illumination. For two hours Peter observed the shading changing in shape and appearance, and he believes that it may have been a haze as a result of outgassing in the area.*

Occultations

There is another facet of lunar work which can actually help contribute to knowledge, and that is the observation of *lunar occultations*, when the Moon passes in front of a star. Basically, this means timing the exact instant at which the star disappears or reappears. This may seem trivial, but it's both more difficult than it sounds and a lunar occultation is a much rarer event than you might imagine. There are surprisingly few stars in the path of the Moon that are bright enough to be seen against its glare – usually about a dozen each month are observable from any one location. Most of them are quite faint and need good equipment (though not necessarily large apertures) and very clear skies to be seen at all.

Your occultation timings must be highly accurate, and your reaction time will be a major source of error. Ideally, your telescope should have a long *f*-ratio and give good contrast, so a 75 mm (3-inch) refractor may be better than a larger Schmidt-Cassegrain (see Chapter 5). A digital watch with a stopwatch facility is adequate for timings, but you must check it before and after the event against an official time signal, such as a radio time signal or GPS unit – Internet Time is probably not reliable enough. Casual timings are of no use.

It's hard to see any but the brightest stars at the bright limb of the Moon with certainty. Most observations are therefore of occultation events taking place at the dark limb of the Moon – that is, disappearances well before full Moon and reappearances well after full Moon. The good news is that you can usually observe lunar occultations just as well from the city as from the country. Wherever you live, the chief requirement is a good clear sky. The light from the Moon itself is the major contributor to the sky brightness.

As in most other fields of astronomy, these observations can now be carried out using imaging equipment. What you need is a low-light video camera – webcam-type devices are not usually sensitive enough.

How do you find out when occultations are due to take place? The Moon's track as seen in the sky depends on the observer's location, so predictions of occultations are published for each country by the International Occultation Timing Association. UK observers can find a list of predictions at www.popastro.com/occultation. For Britain, Australia and New Zealand, a brief list appears in the annual *Handbook* of the British Astronomical Association (BAA), while the annual *Observer's Handbook* of the Royal Astronomical Society of Canada lists events for the US and Canada. These organizations also collect observations and relay them to the International Lunar Occultation Center in Japan. The results are used to refine our knowledge of the orbits of both the Moon and Earth, and are of genuine scientific value.

Occasionally, a star may not be completely hidden during an occultation, but the Moon's upper or lower limb just skims past it. This is known as a *grazing occultation*, and is particularly useful and interesting to observe. The star may appear to wink off and on several times as it is glimpsed through valleys on the limb. You need to be in just the right place to see this happen, and it usually means traveling some distance from your home. The Moon's path in space is not known precisely, so there is usually some uncertainty about the track on the Earth's surface along which the Moon will just be seen to graze the star. Ideally, there should be several observers spread out a mile or so either side of the predicted graze line, at right angles to it. Some observers will probably see no occultation at all, while others will see a brief occultation. Some will be lucky enough to see the graze. Portable telescopes are quite adequate.

Mars

Of all the planets, Saturn notwithstanding, Mars is the one that fires the imagination. With its sandy deserts and thin atmosphere, it is the only known place beyond the Earth where humans might feel even remotely at home. Even through the telescope Mars is usually disappointingly far away – it is a small world, and comes within good observation distance for only a few months every two years or so at the time of opposition (that is, when its orbit brings it closest to Earth). Mars has a very elliptical orbit, and its distance from the Earth at opposition varies by a factor of 1.8. The net result is that the planet is best placed for amateur astronomers only at intervals of 15 and 17 years.

At a favorable opposition even a small telescope will show some detail on Mars, as Figure 4.10 shows, though you really need a telescope of at least 150 mm (6 inches) aperture to see anything worthwhile. Your first glimpse could be disappointing. If the seeing is not very good, you may well see two or three Marses, all shimmering and overlapping.

▶ Fig. 4.10 Mars, photographed through a Meade ETX 125 mm Maksutov by the author using a ToUcam webcam when the planet was 16 arc seconds in diameter. It can become considerably larger than this for a few weeks on either side of each opposition every two years.

Once things have settled down, you will at first see a small orange disk and little else, but persevere and you may notice a distinctly darker shading, and maybe a hint of a white polar cap. You may find that from time to time the seeing improves sufficiently for you to see quite a bit of detail. If you have been observing for some time, perhaps making a drawing, you begin to realize that the markings are no longer in the same places as when you first observed them. Mars turns on its axis at much the same rate as Earth – in 24 hours 37 minutes, in fact – so bringing new features into view. If you observe at about the same time the next night, you will see the same part of the planet.

There is quite a range of features to be seen on Mars: dark and bright areas, polar caps (of which only one is usually visible at any time), and clouds of varying appearance. Strong winds blow through the thin atmosphere, whipping up the dust that covers much of the surface and spreading it around. As the underlying rocks are covered and uncovered, the appearance of the markings we see from Earth changes from opposition to opposition, and sometimes from week to week. Such dust storms can blow up with no warning. I remember taking a look at Mars from an observatory on top of a college building in the middle of Manchester in 1971. I was rather disappointed with my view through the 150 mm (6-inch) refractor, as I could see no detail at all. Shortly afterward I heard that a major dust storm had just begun, coincidentally just as the Mariner 9 space probe entered orbit to begin photographing the planet. Despite my initial impression, my city location had been no hindrance to my observation of the planet.

It is the shifting sands of Mars and the planet's constantly varying appearance that make it so fascinating to watch. As it turns at roughly the same rate as Earth, observations you make could complement those made by Mars observers in other parts of the world, who will be seeing a different part of the planet at a different time. You could be the first to catch the start of a new dust storm. The way the planet changes its appearance is of interest to planetologists who are studying the climate of Mars. After all, one day humans will certainly set foot on Mars, and amateur observations made now will add to the knowledge of the sort of weather they can expect. Even though there are spacecraft constantly orbiting and photographing the planet, they don't have the global view that we get from Earth, so don't assume that NASA has it all taped!

Dust storms were once thought to occur mostly when Mars is at its closest oppositions, when the planet is also closest to and heated most strongly by the Sun. But analysis of observations made since 1873 has shown that they can occur at almost any time, with only brief dust-free spells.

▶ Fig. 4.11 Changes in the size of Mars during the 2005–6 opposition, photographed by Damian Peach. The first image, made in April 2005 when the planet was only 6.4 arc seconds across, is at top left, and as you follow the spiral the planet gets nearer and larger until opposition in November 2005, when it was nearly 20 arc seconds across. Then it receded, until the last observation at center, in April 2006, when it was just 5.0 arc seconds across.

Mars is a notoriously difficult planet to observe away from the times of opposition, because its diameter is very small: only about 4 arc seconds when on the far side of its orbit from Earth. Nevertheless, amateur observers are now able to make worthwhile observations even when the planet is small, using webcam-type cameras (Figure 4.11). For many, this is the preferred means of viewing Mars and indeed any other planet. Amateurs using commercially available equipment can take images of top quality, even from light-polluted urban locations.

Jupiter

Whereas Mars is small and infrequently well seen, Jupiter is a regular sight in our skies and is big enough to show detail in the smallest of telescopes. The planet's brightness means that Jupiter observers can study their target in light-polluted and murky skies that would make other observers pack up. Like Mars, Jupiter presents us with ever-varying features, and its appearance can change significantly from year to year or even month to month. But while Mars is a rocky planet, Jupiter is a gas giant with no observable solid surface. What we see is the changing patterns in its thick upper atmosphere. There are dark *belts*, fascinating to study as they can suddenly fade, causing a radical change in the appearance of the planet. Separating the belts are the lighter *zones*. The Jupiter observer must get to grips with the terminology of these belts and zones, which have such tortuous descriptions as North North Temperate Belt (South).

Study Jupiter and you will see that the edges of the belts are not nearly as straight as they appear at first glance. Swirls of gas lap into the neighboring zones, and white spots and ovals come and go. There is a special nomenclature for these features: Jupiter observers speak of such things as "projections," "festoons," and "barges." Color is immediately apparent – mostly shades of cream and yellow, but also browns, reds and even blues. Most striking is the famous Great Red Spot, which has probably been a permanent feature ever since Robert Hooke first reported seeing such a spot in 1664. It changes its color over the years, sometimes becoming so pale as to be virtually white. More often it is a salmon pink, and sometimes it takes on a true red hue so as to justify its name. Since 2006 it has had a smaller rival, officially known as Oval BA but nicknamed "Red Spot Junior." Curiously, it began life as three white spots that merged in 2000. There are other small red spots as well.

Jupiter rotates very rapidly – in less than 10 hours – so not only do the features move round the disk quite noticeably as you observe, but the planet is clearly flattened. As Jupiter is not a solid body, its rotation period varies with latitude, with a difference of about 5 minutes a day between low and high latitudes. The rapid rotation means that, as with Mars, the planet presents quite a different face from that seen by an observer in another country a few hours later or earlier. Space probes and the Hubble Space Telescope may have shown us much more detail than is visible from the Earth's surface, but they cannot always provide the round-the-clock monitoring that amateur observations can.

The ever-changing disk of Jupiter presents observers with a unique challenge. Minor features come and go, and also drift in longitude, borne by atmospheric currents. Even the belts and zones have occasionally undergone changes. The problem is to record as much as possible sufficiently quickly before the planet's rotation takes some features away and brings others into view. A straight-forward drawing of the planet was once the standard method of

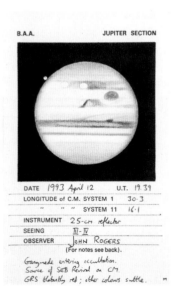

B.A.A.   JUPITER SECTION

DATE	1993 April 12	U.T. 19.39
LONGITUDE of C.M. SYSTEM 1		30.3
" " " SYSTEM 11		16.1
INSTRUMENT	25-cm reflector	
SEEING	III-IV	
OBSERVER	John Rogers	

(For notes see back).

Ganymede entering occultation.
Source of SEB Revival on CM.
GRS blatantly red; other colours subtle.

◀ Fig. 4.12 A drawing of Jupiter by John Rogers of Cambridge, UK, prepared on a standard blank supplied by the British Astronomical Association. It is not necessary to use color when drawing a planet, though in the case of Jupiter there are various hues to record, of the Great Red Spot in particular.

▶ *Fig. 4.13 Simon Kidd of Welwyn, Hertfordshire, UK, took this image of Jupiter on November 29, 2012, using a C14 355 mm telescope and a Flea 3 webcam.*

recording its appearance, and it is still preferred by some observers. The drawing need not be artistically pleasing as long as the features are accurately plotted. A careful record of the view through a 150 mm or 200 mm (6- or 8-inch) telescope can show a wealth of detail. Experienced observers can produce drawings very quickly, and with time the ability to reproduce fine shadings improves. Because the planet's disk is so flattened, Jupiter observers use prepared blanks of the right shape. The drawing shown in Figure 4.12 was prepared on such a blank.

The traditional way of measuring accurate positions for features is to estimate the time when they cross Jupiter's centerline – the central meridian – which is fairly easy to visualize because of the planet's flattening. Timings to the nearest minute correspond to an accuracy of 0.6° of longitude on Jupiter. Such measures can then be used to produce charts showing the changes in longitude of features that result from the different rotation rates and motion in the atmosphere. For example, it is quite common for spots or ovals to change their rotation speed suddenly.

Visual observing has now largely given way to imaging for serious study of the changing face of the planet (Figure 4.13). Further details of the techniques are given in Chapter 6. As a result, amateurs are now able to record features that would not have been noticed in the past.

An example of this happened in 2009, when Australian amateur astronomer Anthony Wesley noticed that a small black spot had suddenly appeared on Jupiter. He was immediately reminded of a much larger series of black marks that appeared in 1994 as a result of the impact of a small comet, Shoemaker-Levy 9, on Jupiter. Those marks had been visible with small instruments, but Wesley's 2009 mark was much smaller and would probably have been missed by visual observers, or at least not appreciated as another impact scar. In due course the feature was intensively studied by the Hubble Space Telescope and other instruments, but it's unlikely that it would have been spotted had it not been for Wesley's image of the planet. You might imagine that Jupiter is constantly monitored by professional instruments, but this is very far from the truth. Time on large telescopes such as Hubble is at a premium, and it is rarely used to observe the planets.

Wesley recorded another impact in 2010, this time a flash rather than a scar, which was confirmed by another video by Christopher Go from Cebu City in the Philippines, and further flashes have since been imaged from city sites. So Jupiter is an object which can hold surprises for everyone, and a light-polluted environment is no hindrance to observing it.

The smallest telescope or even binoculars will reveal the four largest moons of Jupiter, known as the Galilean satellites as they were first reported by Galileo. The constant shuttling of these moons – among the largest in the Solar System – is a perennial fascination. It is hard to see them as disks unless you have a large telescope and good seeing, but the shadows they cast on Jupiter as they cross in front of it are easily visible. The moons can also be seen going into or coming out of eclipse. Amateurs with large instruments can image details on these satellites.

Saturn

Were it not for its glorious ring system, Saturn would be a very neglected planet. It is smaller and more distant than Jupiter, and its disk shows very much less detail. Until it became possible to take high-resolution images of the planet, the most noticeable event was the outbreak of a white spot every 30 years or so, though with the right techniques more detail is visible. But the rings turn a distinctly saturnine planet into one of the true spectacles of the sky. Any star party where Saturn is on show is guaranteed to be a success.

The dedicated Saturn observer, however, must search for more elusive phenomena. The planet's features are more subtle than those of Jupiter, but there is still much to be seen and done, though as with the other planets, the larger your telescope, the better. Like Jupiter, Saturn has dark belts, and these vary in position and brightness. Some small white spots are occasionally seen, and observers also make notes of the colors of the belts and zones. Color filters can help to emphasize any slight color differences.

As with Jupiter, drawings of Saturn are best made on a prepared blank. The problem here is that because of Saturn's high inclination to the ecliptic, the angle at which the rings are tilted toward us changes continuously in the course of the planet's 29½-year orbit, so a single blank will not suffice (see Figure 4.14). Suitable forms may be downloaded from http://www.britastro.org/saturn/visrep.html.

▶ *Fig. 4.14 This sequence of photos taken by Damian Peach between 2002 and 2008 shows how Saturn changes its aspect from year to year.*

▼ *Fig. 4.15 Saturn, drawn by Matthew Boulton of Redditch, Worcestershire, UK, in August 1991. He used a 157 mm (6¼-inch) reflector.*

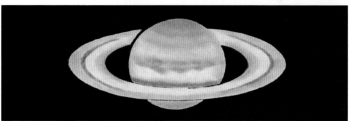

Figure 4.15 shows a typical observer's sketch of the planet. Every 15 years or so the rings are presented to us edge-on, as in 2009 and 2025, and it is at these times that it becomes obvious how thin they really are, since they can virtually disappear. At around these times too we can catch a rare glimpse of the unilluminated side of the rings if the Sun is on the opposite side of the plane of the rings from the Earth.

Any telescope larger than about 60 mm (2.4 inches) aperture will show the division between the outer, narrower, and fainter Ring A and the inner, broader, and brighter Ring B. It is known as Cassini's Division after Giovanni Cassini, who discovered it in 1675. When the seeing is really good you might be able to pick out Encke's Division, within Ring A, first seen in 1837 by Johann Encke. Occasionally, observers report suggestions of additional divisions within the rings or faint rings inside or outside the main ones. The shadows cast by the rings on the globe and by the globe on the rings are also interesting to watch, and sometimes irregularities can be seen. To see details such as these you will have to study the planet carefully – rarely does Saturn reveal its secrets to the casual spectator.

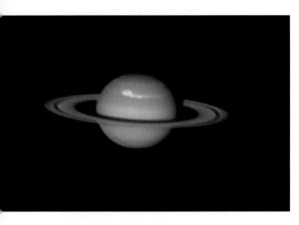

◀ *Fig. 4.16 Saturn, photographed on December 22, 2010, through a 14½-inch telescope from Murrumbateman, NSW, Australia. A white spot is visible in the planet's northern hemisphere, caused by the largest storm seen on the planet for many years.*

An urban location is no handicap to the Saturn observer. The former Director of the BAA's Saturn Section, Alan Heath, lives in a suburb sandwiched between the cities of Nottingham and Derby where the naked-eye limiting magnitude (that is, the faintest star visible) is rarely better than 4. He finds, paradoxically, that very subtle detail, such as Encke's Division, can actually be easier to see when the sky is bright, in twilight. This may be because there is then less glare from the planet's disk. Alan's location is made even worse by a large power station a few miles to the south. When the planet is low in that direction, the seeing can be very bad. Alan finds that his best seeing is in what he calls "pre-fog" conditions, about half an hour before a mist descends and reduces the transparency.

With the advent of digital imaging, Saturn observing came of age. Details previously only suspected by visual observers became easily visible on images taken around the world. Despite the presence of the Cassini probe in orbit around the planet, in 2010 Anthony Wesley (who first detected impact scars on Jupiter – see page 65) was the first to notice a new storm on the planet, which rapidly developed into a major white spot. The spacecraft had been over the night side of the planet at the time (Figure 4.16).

Images also reveal color which eludes visual observers. The hemisphere which points away from the Sun is seen to have a noticeable blue tinge, and amateur images can track subtle color changes even when no spacecraft is sending back images.

The other planets

Mars, Jupiter and Saturn are undoubtedly the main planetary attractions for amateur observers, but the inner planets and those farthest from the Sun are not without interest. For all but Venus, though, you need quite a big telescope to do anything more than track their movements.

Mercury and Venus

Mercury and Venus are both closer to the Sun than the Earth is, so are always close to the Sun in the sky. Elusive Mercury is never more than 28° from the Sun, and is therefore usually seen low down in a light dawn or dusk sky. Most amateur astronomers regard it as a curiosity rather than an object for serious study. To be honest, it serves mostly as a test of your observing skill, for you will do well to make out any markings on it at all visually, and even the imagers have problems, though the markings have been successfully recorded (Figure 4.17). For the urban observer, Mercury may be a particular challenge if there are local obstructions to the east or west. The best conditions for observing it through the telescope are while the sky is still light, but take great care when searching for it that you do not encounter the Sun by accident.

Venus, however, is much more easily seen. As the evening star it is one of the most obvious of planets, hanging like a searchlight in the sky. This is not the best time to observe Venus, for it is simply too dazzling, and usually too low in the sky for good seeing. Like Mercury, it is best observed while the sky is still light.

But unlike Mercury, there are things about Venus that make it worth observing. Although its surface is hidden beneath a dense cloud cover, there are faint shadings which require clear skies and good, clean optics if they are to be glimpsed. Particularly interesting is the time of *dichotomy* – that is, when the planet is exactly at half phase, with the terminator running straight up and down. Whatever sort of telescope they are using, observers usually find that at the time

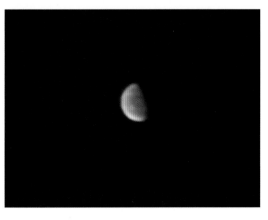

▶ *Fig. 4.17 The late Erwin van der Velden of Brisbane, Australia, took this webcam image of Mercury showing a remarkable amount of detail using a 200 mm Schmidt-Cassegrain telescope.*

when the planet should in theory be at half phase, slightly less than half of its disk is illuminated. Another mystery is the *Ashen Light* – a faint glow on the unlit portion of the crescent Venus, like earthshine on the crescent Moon. Earthshine is caused by sunlight reflected from the Earth and on to the Moon, but there is nothing that can direct sunlight on to the dark side of Venus.

The cause of both the anomalous time of dichotomy and the Ashen Light probably lies in Venus's dense atmosphere. When Venus passed in front of a bright star in 1981, I and a few others who were observing the event from the center of the occultation line saw that a faint image of the star clung to the limb of the planet for a few minutes after the time when it should have passed out of sight. The starlight was presumably being refracted around the limb. If this can happen, then it is certainly possible for sunlight to be refracted around the planet to the dark side, which would offer an explanation of the Ashen Light. Another theory is that the Ashen Light is caused by the release of energy stored in gas molecules during the daytime. But the curious thing is that no images have shown any sign of the Ashen Light, other than at far-infrared wavelengths well beyond the eye's capabilities.

Since Venus observers generally find that they get the best seeing when the planet is high in the sky during daylight, even the most light-polluted urban environment is no hindrance when you are looking for shadings on the planet's disk. The chief requirement is a clear sky. The Ashen Light does need a dark background to be seen, so you are less likely than your country cousins to see it.

Imagers get the strongest markings by using ultraviolet filters, as shown in Figure 4.18.

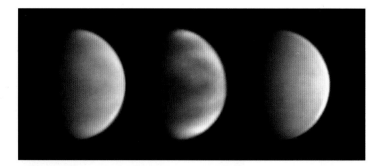

▲ Fig. 4.18 Venus photographed by Damian Peach through a 355 mm telescope on April 20, 2007. The first image is a color composite made from the two mono images shown, the center one being in ultraviolet light and the right-hand one through an infrared filter transmitting at about 1 micron.

Uranus, Neptune and Pluto

Not surprisingly, the far outer planets are harder to find than the others because they are so faint. You need a star map with their positions marked or a Go To telescope to find Uranus or Neptune. Since 2006 Pluto is no longer defined as a planet but many people still regard it as such, even if only because it's the most distant Solar System object that is usually observable. With Pluto you need a fairly large telescope, an accurate position, and observations over several nights to be certain that you have actually seen it.

Uranus and Neptune show tiny disks which are usually feature-less. I've seen the suspicion of a marking on Uranus with a 450 mm (18-inch) refractor from the London suburb of Mill Hill, so I know it can be done. It's always worth looking if you have a large enough instrument. Images made by amateurs under good conditions also show these markings. As for Pluto, you need the Hubble Space Telescope to see even a disk, so just be happy if you find it! Plot the positions of stars in its vicinity from night to night, and you should find that Pluto has moved. It is around 14th magnitude, so you need at least a 200 mm (8-inch) telescope to have any chance of tracking it down visually, though you can take an image of it with a smaller telescope.

There is a trick to use when looking for objects near the limiting magnitude of your instrument. If the seeing is good, you can mag-nify the image up to the limit that your telescope will bear without making star images any fainter, since they are points of light. The sky background, however, gets darker as the magnification increases since its light is being spread over a larger area. So you can increase the brightness difference between the object and the background by using high powers.

Also, faint objects are best seen if you do not look directly at them. Your eye is less sensitive to light right at the center of the field of view, although that is where it gives the best resolving power. So use *averted vision* – that is, look a little to the side of where you expect the object to be. To avoid the faint image falling right on your blind spot, keep the object to the side of the field of view nearest your nose, whichever eye you are using.

You may find that making the telescope quiver slightly helps you to be sure that what you can see is a star rather than a defect in your eye. Looking hard can help you to see clearly. It has been claimed that light builds up in your eye over a matter of seconds, just as it does on an imaging device, though there is no physiological evidence for this. I suspect that it is the result of concentration and avoiding the small eye movements we unconsciously make.

Asteroids

These minor planets range in brightness from about magnitude 5.5 for Vesta to as faint as you can go. They look just like stars ("asteroid" means "starlike"), except that they change position slightly from night to night. For most of the 20th century, asteroids were regarded as little more than a curiosity, mostly orbiting placidly way out in the Solar System between Mars and Jupiter. But interest in asteroids has increased in recent years, among both amateurs and professionals, with the realization that some of them have orbits that can bring them close to Earth, and the potential in the future actually to hit the planet and possibly cause major devastation. From a more academic point of view, some of them probably represent the basic material from which the Solar System was formed.

While most Near Earth Objects (NEOs) are likely to be discovered by intensive professional searches, amateurs can play an role in measuring the positions of newly discovered objects, though these are usually rather faint. This is work exclusively carried out using CCD cameras though not necessarily very large telescopes. Sometimes an NEO actually does come close, maybe with little warning, and on those occasions it can be comparatively bright but fast-moving. With skill you can view it directly, moving perceptibly through the field of view of the telescope or maybe even binoculars.

Other useful tasks that can be carried out using the powerful combination of well-mounted telescope and CCD or low-light video camera are to measure the brightness and color of asteroids as they rotate, and to study the surface properties of asteroids. Even though they are usually too small and remote to show any disk, there is much to be learned from simply measuring how their brightness changes as they rotate and with the Sun shining on them at different angles.

Predictions of bright asteroid positions are published in the *BAA Handbook* each year, and also in the RASC *Observer's Handbook*. Smartphones with sky apps may be able to display the precise position of any chosen asteroid on their map, or you can go to ssd.jpl.nasa.gov/sbdb.cgi#top and enter the asteroid's name or number in the search box. This is particularly useful for getting positions for NEOs, as you need predictions (technically known as an *ephemeris*) for your own location.

Comets

Many amateur astronomers, particularly those who live in cities, would unhesitatingly vote comets the most frustrating of astronomical objects, for several reasons. First, they are potentially the most spectacular of celestial sights. Second, they can be extremely fickle, arriving on the scene with great promise yet delivering virtually nothing. Third, they have a habit of disappearing in the slightest hint of light pollution. While from the city we strain to see a hazy head, from a dark country site the same comet clearly sports a magnificent long tail. There is a saying: "Comets are like cats. They have tails and do exactly as they please."

Living in the city is little hindrance to observing the objects described in this chapter so far. But when it comes to comets, the streetlights and pollution can easily beat us. I'm not talking here about the rare great comet, which is so bright that it can be seen even from city centers, such as Comet Hale-Bopp in 1996 or Comet McNaught in 2007, but the run-of-the-mill comets that come and go every year and provide interest for those in darker skies. Usually there are several of these, with one every year or two that is bright enough to be seen with the naked eye under the right circumstances.

The problem with comets is that they often have a low surface brightness, even though they may span several degrees of sky. Only the immediate surroundings of the head are reasonably bright, and this is all we see in our light-polluted city sky. Even a comet that reaches its predicted brightness may appear much fainter. Published magnitude predictions refer to the total amount of light from the comet and may well imply naked-eye visibility, such as magnitude 4 or 5. But much of that brightness is spread out over a considerable area. The head of the comet is much fainter, perhaps about magnitude 7 or 8. Comets often appear in twilight, low in the sky, since they brighten up only when they are near the Sun. Under these circumstances, the urban astronomer has the same break as someone in the country, as the twilight is the same issue for both of them. Figure 4.19 shows a comet that was just visible to the naked eye in twilight from both the city and the country.

Some other comets are condensed, and have the courtesy to travel high in the sky, so may be visible in urban skies. You'll need good transparency, a well-baffled telescope (see page 117), and a site shielded from any nearby streetlights. To find the comet most observers these days will use a computer sky map and enter the comet's orbital details, or go to the excellent website www.heavens-above.com which lists the brighter comets and displays them on a map. The SPA/BAA joint comet site at www.popastro.com also gives positions for comets as well as finder charts for the brighter ones.

A few moderately bright comets appear most years, but rarely do they meet any of these requirements. By "moderately bright" I mean better than ninth magnitude. Jonathan Shanklin, Director of the SPA and BAA Comet Sections, lives near the center of Cambridge, England, a city with a population of about 100,000. He can observe

◀ *Fig. 4.19 Comets are often visible in twilight, at a time when both urban and country astronomers are in the same boat. This is Comet 2011 L4 (Panstarrs), photographed from Edgware, London, on March 29, 2013, using a 355 mm telescope.*

well-placed comets down to ninth magnitude from his home using 20 × 80 binoculars. But the city comet-hunter must be patient, and be prepared for a long wait before a suitable object appears.

This dismal picture has been changed completely by the arrival of CCDs. These electronic detectors have a remarkable ability to slice through light pollution, revealing comets that even country observers with large telescopes cannot see (see Chapter 6).

Visual estimates of a comet's brightness are useful, but urban conditions will probably result in inaccurate results, for the reasons given above. Should the comet be bright enough to show a tail, however, another feature to look for is a *disconnection event*, in which a break appears in a comet's gas tail. Comets generally have two tails – a dust tail, which is the one best seen visually, and a gas tail, which points directly away from the Sun. Observations of disconnection events can yield information about the solar wind of particles streaming away from the Sun.

Very, very occasionally there is a comet which throws modesty to the winds and reveals itself in all its glory, tail and all. Comet Bennett in 1970 was such a comet. It hung like a dagger in the dark morning sky. Shift workers coming off duty were alarmed to see it, as it was poorly publicized and they had no idea what it might be. I could understand why comets were regarded as portents in earlier times. The astronomy publicity machine is much better now than it was in 1970, and the appearance of Comet Hale-Bopp in 1997 was widely enjoyed by both astronomers and the public. A really bright comet can be seen despite light pollution. Some are visible for only a few days in the bright twilight, and may not get the publicity that longer-lived objects enjoy. One such was Comet McNaught (2006 P1), which from the northern hemisphere was seen only briefly in the evening twilight in January 2007, though it put on a much better and longer show in the southern hemisphere.

When comets first appear they are usually dim fuzzy patches, indistinguishable in a small telescope from such objects as nebulae and galaxies. Hunting for new, undiscovered comets is a dedicated pursuit and requires good sky conditions. Few comets are discovered visually these days. It is most unlikely that a city observer will be the first to spot a comet. There is just a chance, however, that a new great comet will appear first in the twilight sky, having crept up unnoticed from behind the Sun. In this case, the first person to see it could just be some high-rise city astronomer gazing at the distant horizon. Maybe the big one is just around the corner, and the skies of our cities will once more be graced by a spectacular sight that will make everyone, briefly, a skywatcher.

Meteors

Out in the country, the peak period of a meteor shower is a delight to watch. Every few minutes it seems as if a star has broken away from its moorings and dashed with a fiery trail toward the Earth, justifying the popular name of "shooting star." We know that they are really just tiny flakes of dust from comets' tails that burn up in the atmosphere, but watching for them is still great fun, and can actually be useful. (A point of nomenclature that it's best to make clear: the particle that burns up is the *meteoroid*, the fleeting trail of light we see is the *meteor*. A *meteorite* is a larger, more substantial object that penetrates to the Earth's surface.)

Meteor watching is certainly best suited to those clear, dark skies we all dream of. Simple observations of the brightness and time of meteors, correctly recorded and corrected for observational errors, can be of use in charting any variations in the density of the stream of meteoroids which the Earth is encountering. Meteoroids that give rise to a particular shower have their own orbit around the Sun, and the shower occurs at the particular time of year when the Earth traveling in its orbit crosses the orbit of those meteoroids. For some showers the meteoroids are clumped in their orbits, leading to higher rates in some years. The range of meteoroid sizes also varies from shower to shower, as revealed by the brightness of meteors observed.

From city skies, however, it is all very different. Instead of a meteor every few minutes, we may be lucky to see one every half hour or even every hour. Dispirited, we give up and curse the handbook that suggested an hourly rate of 60 or 100. The problem is partly that the sky is too bright, but also that the quoted rates apply only under ideal conditions, which rarely apply, even in the country.

Because we are probably not looking at exactly the spot where a meteor appears, and because it is a rapidly moving, momentary flash and not a steady source of light, we cannot see meteors anything like as faint as we can stars. In a dark country sky, with a limiting magnitude of 6.0, we would not expect to see meteors fainter than magnitude 4.5, or maybe 5.0 if we are lucky. Even then, we certainly don't see all the fourth-magnitude meteors that appear in the sky. In a city sky, where the limiting magnitude may be only 4.0 or worse, we are restricted to meteors brighter than third magnitude. The numbers of meteors we see usually increases sharply with their faintness: there are actually about two or three times as many fourth-magnitude meteors as third-magnitude ones. So by cutting out a whole magnitude, we are considerably reducing the rate we can record.

Having said that, it is interesting to look at the brightness distribution of meteors actually observed in dark skies. Around 60% of the meteors seen in good showers such as the Perseids or Quadrantids are of second magnitude or brighter, and are therefore easily observed even in average city skies. The problem is partly a psychological one – it is hard to sustain the enthusiasm for observing meteors when very few are visible. Even under good skies on the night of a shower's peak activity, there can be long intervals between meteors for a single observer who is viewing the whole sky.

Forecasts of meteor numbers for a particular shower are given as the *zenithal hourly rate*, or ZHR. This is the number of meteors a skilled observer would record under a perfect whole sky of limiting magnitude 6.5 with the radiant at the zenith. The *radiant* is the point from which all the meteors of a particular shower appear to radiate, and it is very rarely in the zenith. If it is an altitude of 45°, only about 70% of the ZHR would be recorded – still with perfect sky and perfect observer. If your sky and you are less than perfect, you will see only some of the brighter meteors, and the rate drops further. If you are observing from the city you probably have a restricted horizon, which may cut out another 15% or 20%. It is no wonder that you see so few.

So don't be put off looking for meteors, particularly on a shower night, but you will have to lower your expectations even more than those in the country. And there is always the chance of observing a really fantastic meteor, or even a sudden burst of meteors that no one was expecting. These do happen from time to time, and can catch everybody unawares.

Lights in the sky

Under this catch-all heading are included various natural phenomena of the upper and lower atmosphere, from aurorae to sun pillars, and miscellaneous topics such as observing artificial Earth satellites.

Aurorae

The northern (or, in the southern hemisphere, southern) lights are another phenomenon that many urban astronomers just dream about seeing. While in polar latitudes they are a common feature of the long winter nights, for most observers in Britain they are a rarity. The lower the latitude – more correctly, the farther from the magnetic pole – the less likely they are to occur. The north magnetic pole is located in northern Canada, so observers in Canada and the northern US are much better placed for seeing aurorae than their counterparts in the densely populated parts of Europe. They are more common when the Sun is spotty rather than at solar minimum.

Faint aurorae occur fairly frequently low in the northern skies of Britain and the northern US, but they often go unnoticed since you need a perfectly dark sky and clear horizon to see them. Although the predominant color of an aurora is generally green, from lower latitudes we see only the upper part of the display, which more often than not is red. From time to time there is an aurora bright enough even for observers near cities to be aware of it (see Figure 4.20). The sky can become very green or red, and various distinct patterns, given names such as rays and bands, can develop. The event is better seen from dark skies, but always bear in mind that an aurora may appear at any time. Every time you go out to observe, glance at your northern horizon. You will get to know how bright it normally is, and should be able to detect any change. One trick is to take a quick wide-angle photo using a digital camera, with an exposure time of maybe 15 seconds at a high ISO rating. You don't even need to support the camera on a tripod but simply hold the camera as steady as possible. Any aurora will show up red or green, rather than the usual light-pollution color.

If an aurora is in progress, it may last only for a few more

▶ *Fig. 4.20 Occasionally aurorae are bright enough to be seen in suburban skies. This one, in March 1989, photographed from the author's home in Greater London, was strong enough to overcome not only streetlights but a first-quarter Moon.*

minutes, or it may continue for hours. You could jump in the car and head for your darkest local site, but by the time you get there the display could well be over. So take some contingency photos with the camera on a tripod. (There is more on sky photography from urban areas in Chapter 6.)

Noctilucent clouds

Aurorae and meteors are both phenomena of the upper atmosphere. A third, less well known but very beautiful and definitely visible from some cities at the right time, is noctilucent clouds. These fine, wispy clouds appear in the summer months in the northern sky. They are more common at latitudes between 55° and 60°, but from time to time they are seen as far down as latitude 50°. This makes them a feature of northern European and Canadian skies, rather than American skies. You may think that fine wispy clouds are nothing unusual, but noctilucent clouds become visible long after the Sun has set. They have a characteristic silvery-blue color that is brought out well on color photos, as in Figure 4.21. Noctilucent clouds are quite bright and are visible in the late twilight sky, before light pollution has got much of a grip, so they can be seen quite well from cities.

▲ Fig. 4.21 Noctilucent clouds are delicate clouds which form high in the Earth's atmosphere, at five times the altitude of the highest normal clouds. They are visible only from certain latitudes, but are bright enough to be seen in city skies. These were photographed by Alex Lloyd-Ribeiro from Durham, northern England, in June 2006.

Noctilucent clouds are becoming more frequent, for reasons which are not understood. They form when water vapor condenses on particles at the low temperatures that prevail at altitudes of around 82 km (51 miles). It is not certain what these particles are, but it may be that industrial pollution or the increase in air transport has provided more high-altitude particles around which these clouds can form. Anyone in the right part of the world should keep an eye on the northern horizon after twilight during the summer months for these delicate and beautiful phenomena.

The zodiacal light

While on the subject of occasional lights in the sky, I must mention the zodiacal light, if only for the sake of completeness. This is a pale cone of light that appears after sunset and before sunrise along the line of the ecliptic, the path followed by the Sun, Moon and planets across the sky. It's caused by dust particles way out in space, in the inner Solar System. The best time to see it is when the ecliptic is at its steepest angle to the horizon, which is in spring in the evening sky and in the fall in the morning sky.

You need good, dark, clear skies to see it, as it is chased away by the slightest hint of light pollution. When you do see it from a good location, however, it seems so obvious that you wonder why you missed it before. The reason is that the light is always there, but it is so smeared out by the presence of water vapor and haze that even in the absence of light pollution it is indistinguishable from twilight. Add artificial skyglow and it disappears completely. The zodiacal light is one of those phenomena that urban observers must resign themselves to missing.

Atmospheric phenomena

Although not strictly astronomical, various atmospheric effects are of interest to astronomers. These include mock suns, haloes and sun pillars. A *mock sun* (alternatively called a *sun dog* or *parhelion*) is a bright patch of light, often brightly colored, at the same distance from the horizon as the real Sun but 22° away from it to one side. Quite often you can see two at once, on either side of the Sun. These are most often visible when the Sun is fairly low in the sky) but a solar *halo* can sometimes be seen when the Sun is high up. This takes the form of a pale ring around the Sun, with a radius of, again, 22°. A *sun pillar* (see Figure 4.22) is a bright column above or below the Sun as it rises. All these phenomena are caused by ice crystals suspended in the atmosphere. This does not mean to say that they are restricted to cold weather; at high levels, typically where cirrus clouds form, the temperature is well below freezing even in the tropics.

▲ *Fig. 4.22 This sun pillar was* *though the phenomenon can also*
photographed one wintry morning, *be seen in summer.*

Another atmospheric phenomenon connected with the Sun is the *green flash* – a green upper rim to the Sun's disk briefly visible just before it sets if the atmosphere is particularly stable. If the setting Sun has a jagged edge as a result of layering in the atmosphere, there is quite likely to be a green flash as long as there is not too much haze. It is best seen through binoculars – but beware: use them only if the Sun is not painful to look at with the naked eye. The green flash is caused by refraction of sunlight by the Earth's atmosphere. A clear, low horizon is needed to see it, so this is a sight for those in high-rise blocks.

Identifiable flying objects

Every so often astronomers are called upon to interpret observations of what are generally termed unidentified flying objects, or UFOs. Curiously, I have never met a regular watcher of the sky who has seen any object that was not readily identifiable. The non-astronomical public are largely unaware of the range of phenomena that can be seen in the sky, and this unfamiliarity gives rise to many bizarre reports which turn out to be easily explained. It is always worth knowing the kind of things that give rise to UFO reports.

Venus, Jupiter or Sirius when low in the evening sky account for many UFOs, while weather balloons or aircraft are behind many more. Living not far from London's Heathrow Airport, I am familiar with the sight of a long string of glittering beads hanging in the sky – the headlights of several aircraft lined up on the approach path. These can be seen from up to 50 km (30 miles) away. When the wind is coming from an unusual direction, the planes use a different runway and the string of lights is seen by people who are not familiar with the sight.

Another form of UFO, reported particularly from cities, can sometimes be seen with the naked eye or in binoculars. It consists of a pale yellow or orange blob, making its way steadily across the sky with an undulating motion. This is nothing more than a bird with a white belly, illuminated from below by the streetlights. A brighter version, often seen on special occasions such as New Year's Eve, is the Chinese lantern – a paper hot-air balloon driven by a flame.

Occasionally, even experienced observers get caught out by blobs of light which seem to dance around the sky. These are caused by banks of searchlights, of the type that often accompany an event such as a rock concert or funfair, playing on a thin layer of haze. Events like this are sometimes held in out-of-the-way places, so the lights could appear where you least expect them. Laser light shows produce starlike points which can suddenly dash across the sky.

Artificial satellites

Many people are surprised to be told that artificial Earth satellites are easy to see, and that their sighting of a strange bright light silently crossing the sky is of just such an object. Sometimes you may be misled by reports that the object moved irregularly, but this is an optical illusion. When the only reference points are stars, people can become disoriented and believe that the satellite, whose motion is essentially straight and regular, is winding its way among the stars or moving in jerks.

There are hundreds of faint satellites, and they are mostly seen in summer, when the Sun is not far below the horizon. But the most spectacular are the International Space Station (ISS), which can be one of the brightest objects in the night sky after the Moon, and Iridium flares. The ISS orbits every 90 minutes or so, and covers all parts of the Earth's surface between latitudes 51° north and south, always traveling roughly from west to east. Once seen it is unmistakable, as it is brighter than most planes yet of course silent.

Iridium flares can be just as bright. They are caused by sun glints from polished surfaces on a series of satellites used for phone services, and these glints are highly predictable. You first see a fairly faint satellite which quickly brightens over a matter of seconds and then fades away again. Predictions of both ISS and Iridium flares are given on the website www.heavens-above.com, hosted by the German space agency DLR. You need to choose your precise location, so it is worth registering (free) rather than choosing each time. Again, they are so bright that being in a light-polluted area is no drawback.

Variable stars

If such phenomena as the zodiacal light are beyond the reach of city-dwellers, variable stars are highly accessible. What is more, estimates of variables are among the most scientifically useful observations the amateur can make, and very little equipment is needed.

Most stars are of fixed brightness, but a significant number do vary. With many the changes can be monitored by making careful estimates, either with the naked eye or using a telescope or binoculars. While these estimates may lack the precision of measurements made using electronic devices such as CCDs (see Chapter 6), they have the merit of being easily and quickly made. One of the advantages of variable-star work is that you can make estimates even when only a small part of the sky is clear. While planetary observers need a good view of the ecliptic, there are variables all over the sky. Furthermore, an estimate can be made in a few moments if necessary, making the most of those nights when it may be crystal clear between numerous scudding clouds. And this is one field of measurement where visual observers can still hold their own over the CCD fraternity. While it takes just a few seconds to acquire an image of a variable star, getting an accurate magnitude from it requires care and good technique, so many people still opt to observe visually.

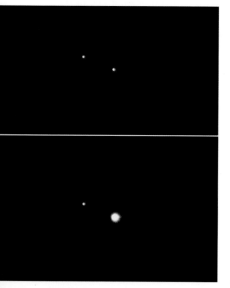

Variable stars come in all varieties. Some vary because one member of a binary system periodically hides the other from our view, causing a dip in their combined brightness. These are called eclipsing binaries. Pulsating stars, such as Cepheids and RR Lyrae stars, vary as their

◀ Fig. 4.23 Light pollution does not prevent you from following the light variations of a variable star using binoculars. Mira (Omicron Ceti) is the right-hand star of the pair in each case, and at the top is near minimum, and on the bottom is at maximum. The left-hand star is unconnected with Mira and is about magnitude 9.5.

▶ *Fig. 4.24 A variable-star chart supplied by the British Astronomical Association's Variable Star Section showing the field of the variable stars R Scuti, S Scuti and V Aquilae, and the magnitudes of comparison stars. The stars in the Scutum area are easily visible with the aid of binoculars from anywhere in the world, from February (in the morning sky) to November (in the evening sky).*

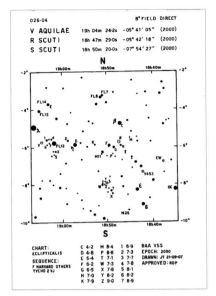

surface area changes. All these categories of variable star have precise periods of variation. Others vary fairly regularly, such as Mira in the constellation of Cetus, which reaches a maximum brightness usually of about magnitude 3 or 4 every 330 days or so, then beginning its fade to about magnitude 9 (see Figure 4.23). Other stars are semiregular, or may be unpredictably irregular in their variations. There are those that suddenly flare or suddenly fade, and those that suddenly cease their variations for a while. Novae and supernovae undergo huge outbursts which increase their brightness by 10 to 20 magnitudes before fading back to obscurity. When one appears, its brightness changes need to be followed closely. With such a motley collection of stars, there is always something happening and individual observers have a real chance of spotting a new development ahead of anybody else.

There are several techniques for making variable-star estimates, but basically you compare the brightness of the variable with that of other stars of known, fixed brightness. These comparison stars should be fairly close in the sky to the variable, ideally within the same field of view of the instrument you are using. Some should be slightly brighter than the variable, and some fainter. If the variable has a wide range of brightness, then many comparison stars are needed to cover the whole of its range. Another requirement for comparison stars is that they should not be strongly colored, since the eye's sensitivity to color varies with brightness.

Good lists of comparison stars for variables are available from organizations such as the American Association of Variable Star Observers (AAVSO) and the BAA Variable Star Section. These lists usually take the form of detailed star charts of the region of the variable, with the comparison stars indicated. An example is shown in Figure 4.24. Some charts mark the comparison stars by letters, and a separate list gives the magnitudes of the comparisons; on others the magnitudes are marked directly to the nearest tenth (but with the decimal point omitted, so that 58, for example, indicates magnitude 5.8). Armed with these, you first locate the area of the variable by using a more general star atlas. Then comes the business of making the estimate itself.

The basic method is to choose two comparison stars which are slightly brighter and fainter than the variable, and then to work out where the variable fits between them. There are two variants. If you are using the BAA's *fractional method*, you make an indirect estimate. If the variable (V) is just slightly fainter than star A and quite a bit brighter than star B, for example, you may decide that it is a quarter of the way from A to B. You would then record your estimate as A(1)V(3)B. Back indoors, you convert this into an actual magnitude. With the AAVSO *interpolation method*, you make a direct magnitude estimate. Knowing your chosen comparison stars to be magnitudes 7.5 and 8.1, say, you picture the difference between them as six 0.1-magnitude steps, and place the variable on this scale. With both methods, the process should be repeated with several pairs of comparison stars, and the results averaged. Most other national observing organizations have adopted one of these two methods.

The human eye is actually quite sensitive to small differences in brightness, and once you have a little experience of making estimates, you can begin to train yourself to recognize magnitude differences using just one comparison star. Then you can use *Pogson's step method*, and estimate the variable's brightness more directly. It may appear 0.4 magnitude brighter than one comparison star, say, so you have an immediate estimate of the variable's brightness. You then repeat the exercise with another comparison, and average the results. If necessary, it is possible by this method to make an estimate using just one comparison star.

The beauty of variable-star observing for the city observer is that, even from a badly light-polluted location, quite faint stars are visible with a minimum of optical aid. The background skyglow may be bright, but its effects can be reduced by using more magnification. John Isles, former Director of the BAA Variable Star Section, recommends using slightly higher-power binoculars from the city than you would otherwise choose. He favors 12 × 40 binoculars for this purpose, though the more common 10 × 50s are also suitable, and even lightweight 8 × 30s for the brighter variable stars, those of magnitude 4 to 6. Although these stars may be visible with the naked eye even from the city, binoculars will make them much more easily observable and will therefore improve the accuracy of the estimates.

John made up to 2,000 variable-star estimates in one year using such binoculars from the roof of a building in London's Covent Garden. Those unfamiliar with London might think from the name that this is an oasis of greenery, but in fact it is in the heart of theaterland and has one of the most light-polluted skies you could imagine. He advises that it is a good idea to mount the binoculars on a tripod. This is especially useful if you want to see stars near the limit of your sky, as city observers are very likely to do. With 12 × 40s he could reach magnitude 9 on a good night, and magnitude 8 on an average night. Image-stabilized binoculars are a more expensive way of doing the same thing.

One word of caution John offers to urban observers of variable stars is that if your sky is noticeably orange as a result of sodium lighting, red variables such as Mira-type stars and some semiregular variables will tend to appear fainter than they really are. It is then best to concentrate on other types of variable. You may be tempted to try to suppress the glow using light-pollution reduction filters (see Chapter 6), but these are frowned on by serious variable-star observers as they may give rise to systematic errors which can vary from star to star. This is because there may be strong features in the spectrum of the star which some filters would cut out. However, the difference is likely to be small, particularly when you are dealing with a star which has a large range of variability.

Variable-star work can also be carried out by telescope, to reach fainter stars than are visible through binoculars. Among the objects that can be monitored in this way are active galactic nuclei. These are the centers of galaxies in which a huge amount of energy is being generated from a very small region, and this energy output can vary irregularly. In many cases the galaxy itself cannot be seen, and certainly not from the city, but the nucleus itself can be located in just the same way as a variable star. Professional astronomers regularly ask the amateur community to keep a particular object under surveillance, and charts are issued giving suitable comparison stars. Such objects are usually fainter than 12th magnitude, so a telescope of fairly large aperture is needed for visual observations.

The army of amateur variable-star observers is mobile, widespread and versatile, and plays an important part in this field of astrophysical research. It is a field where cooperation is very important, and it is more or less essential to join a group such as the American Association of Variable Star Observers or the BAA Variable Star Section, so that you can use the same charts and comparison stars as others.

Nova and supernova patrols

Novae and supernovae are stars which suddenly increase dramatically in brightness. Most observed novae are within our own Galaxy. They undergo an increase of typically 10 to 15 magnitudes, which raises them from the ranks of remote and undistinguished stars in the Milky Way to as much as naked-eye magnitude. Supernovae undergo much more violent outbursts, maybe a staggering 20 magnitudes, which means that for a few days they may outshine all other stars in a galaxy. Supernovae are usually seen in distant galaxies, but there is no reason why one should not appear in our own, except that they are exceedingly rare. The last to be observed in our Galaxy was in 1604, though observers in the southern hemisphere were treated to a supernova in the Large Magellanic Cloud, a small companion galaxy to our own, in 1987. This reached third magnitude, despite the fact that it was about 170,000 light years away. Had it been within our own Galaxy and in a location unobscured by intervening material, it could easily have overtaken Sirius as the brightest star in the sky.

Some dedicated amateurs search for novae and supernovae, and until the 21st century had a fair degree of success. Nova hunters scan the Milky Way, where most novae occur, looking for stars that are not usually there. This can be done visually, which may seem an impossible task, but some people really do know the sky that well. If the idea of learning the position of every star down to ninth magnitude seems daunting, you could always start by getting to know a selected area, then expanding the search. The legendary British observer George Alcock, who discovered six novae and five comets (see page 194), had a photographic knowledge of the sky and could spot interlopers almost as soon as he saw them. Similarly, American observer Peter Collins, discoverer of three novae, spent thousands of hours learning the stars of the Milky Way down to eighth magnitude. Today, CCD or DSLR camera patrols are more likely to be used. In December 2013 the brightest nova for many years was discovered in Centaurus using a DSLR by Australian amateur astronomer and volcano adventurer John Seach, following on just a few months after another bright nova was discovered in Delphinus in a CCD patrol by Japanese amateur Koichi Itagaki.

Nova hunting remains a field where a diligent amateur can still hope to make a discovery, as long as they are prepared to put in many hours of fruitless effort. Guy Hurst is Editor of *The Astronomer* magazine, which is dedicated to listing recent observations and which encourages such patrols. He says, "Modern equipment, including CCDs and DSLRs, are favored over visual because they offer easier comprehensive checking in the comfort of indoors thanks to various items of software." But, he warns, "Fully automated checking does not always seem to be successful and many false alarms occur with spurious objects turning up, sometimes more obvious when looking at raw color images from DSLRs." However, the search could be rewarded by knowing that astronomers all over the world are studying a star which you were the first to spot.

Supernova patrolling has a similar appeal and approach. First you locate a galaxy, then you compare each of the nearby stars with charts or photographs of the area, looking for newcomers. Many visual discoveries of supernovae have been made in this way by the Reverend Robert Evans, observing from country locations in Australia. From urban areas, the job is made that much harder if you cannot see the galaxy in the first place. The supernovae themselves are generally faint – 13th magnitude is considered bright – but objects of this brightness are within the reach of many amateur telescopes, even in city skies. With a CCD and quite a small telescope you can reach fainter objects with exposure times of a few seconds.

So in theory there is no reason why an urban astronomer should not discover either a nova or a supernova. However, in recent years professional sky surveys have become much more efficient and the amateur's advantage of being able to spend time on a job without counting the cost has been taken away by automated systems. A chance remains, but it becomes slimmer and slimmer all the time. And should you spot something that doesn't seem to be on star maps or photos, you first have to eliminate all the other possibilities, from the blindingly obvious such as bright planets down to asteroids, known variable stars, flaws on the imaging chip, ghost reflections inside the telescope and plenty more.

To prove the point that supernovae can still be found from city environments, in January 2014 a supernova was discovered in the bright, nearby galaxy M82 from a site only 14 km (9 miles) from the center of London. It was spotted during a training session at the University of London Observatory in Mill Hill, sandwiched between the M1 and A41 routes out of London. The telescope being used was a Celestron C14, a standard amateur instrument, with CCD camera. Having imaged M82 before clouds rolled over, lecturer Dr Steve Fossey noticed a star which he believed was not usually there. After checking that it was absent on previous photos, and taking an image with another telescope at the observatory in case it was some artifact of the CCD, Fossey alerted other observatories and the supernova was confirmed a few hours later. Amazingly, it was then spotted on photos taken several days prior to Fossey's, but no one had noticed the additional star within the galaxy. At 12th magnitude, it was easily visible with amateur instruments and was the closest supernova to Earth for decades.

Exoplanet monitoring

When I was preparing the first edition of this book in 1994, the very idea of observing planets around other stars – exoplanets – was virtually unheard of. Then in October 1995 I received an early morning phone call from the BBC. Would I go in and talk on breakfast TV about how the first planet had been detected orbiting another star? I raced in, parked in the Director General's spot, and was on air within the hour, having heard enough on the radio about how it was done to get through the interview.

Even more amazing is that it is now possible for amateurs with quite modest equipment to make useful observations of distant exoplanets. The technique is to measure the brightness of their parent star as the planet passes in front of them – what's called a *transit*. The dip in brightness is small but measurable. Many of these stars are comparatively bright, and there is no reason why an urban astronomer should not make the observations. You do need to take great care over the measurements, and this is not a program for the dabbler, but it's another case where light pollution is not a major barrier.

Double stars

Individual stars, on the whole, are rather monotonous. But some otherwise ordinary stars are real beauties because they form part of a double-star system. They are among the showpieces of the sky, and feature high on the urban astronomer's list of favorite objects. While some double stars look close just because they happen to lie in the same line of sight from here on Earth, most are physically associated binary systems – two stars in orbit around each other.

There can be few August star parties where someone is not gazing at Albireo, Beta Cygni (see Figure 4.25). You can have endless discussions about whether its two components are blue and yellow, green and yellow, or just plain white. Even plain white doubles can be attractive to look at, notably where there is more than a straight-forward pair of stars to see, such as in the famous "Double Double," Epsilon Lyrae. There are many other popular doubles in the sky whose appeal lies in their beauty. But there is a serious side to double stars if you want to get more deeply involved.

Measurements of double stars, like those of variable stars, are scientifically useful. The fact that double-star work is unpopular with amateurs shows that scientific usefulness not surprisingly takes back place to other considerations. Most amateur astronomers do

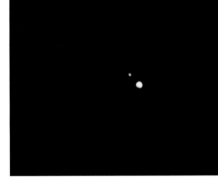

▶ Fig. 4.25 Albireo, or Beta Cygni, is a star which can be seen as a double with any power greater than about 10×. Observers disagree over the colors of the stars, which are emphasized by the contrast between them, but this photograph shows them as blue and yellow.

what they do for fun, and any worthy results are byproducts. To measure double stars visually you need a telescope with a reasonably long focal length and good optics equipped with a *micrometer*. With the micrometer you can measure the *separation* and *position angle* – the relative orientation – of the two stars. The device is fitted in place of the eyepiece.

There are advantages to measuring some double stars visually, particularly where the two components are very dissimilar in brightness, as a CCD tends to bloat the image of the brighter one while the human eye sees them as two points of light.

The lists of double stars which appear in practical observing handbooks, such as *Norton's 2000.0* star atlas and the annual *BAA Handbook*, give position angles and separations, so it might seem that there is little more to be done. But double stars are very important in astrophysics, because they provide a direct way of discovering the masses of individual stars. Since in many binary systems the two stars appear to move around each other very slowly, continued measurements help to refine the orbit and perhaps reveal other companions. The sky is full of double stars, so there is plenty to be done. And, like other stars, they are not seriously affected by skyglow. This is work that can done from the city just as well as from a dark country site. It does, however, demand an equatorially mounted telescope with a focal length long enough to make it easy to obtain high powers.

According to double-star observer Bob Argyle, formerly of the Royal Greenwich Observatory, the minimum aperture you need is about 200 mm (8 inches). He observes from a site about 1.5 km (1 mile) from the center of Cambridge, only 300 meters from a main road lit by high-pressure sodium lights. He finds that with a 200 mm refractor he can make double-star measurements nearly every clear night from this location, and has made more than 9,000 in 20 years.

Quite faint double stars are often neglected, though ones whose components are between 2 and 10 arc seconds apart are easy to separate with even a small telescope. But a moderately large aperture is needed to give good images of the stars rather than just separate them, and Bob regularly uses a magnification of 450 and a micrometer from Cambridge. He emphasizes the need for a solidly mounted telescope with a good drive and slow motions: "At high magnification the slightest bump can move the images considerably and if, in trying to set the [micrometer] wires, you need both hands to continuously move the RA and dec slow motions, then...." He leaves the picture of frustration to the imagination. This requirement probably rules out the commercial Schmidt-Cassegrain telescopes on their standard mountings.

Amateur astronomers have used imaging devices to improve their observations in virtually every other field of observing, so what about double stars? A fairly large image scale is needed – at least 0.5 arc second per pixel (picture element), and the exposures should be made through filters to reduce the tendency for the Earth's atmosphere to act as a prism and smear a star image out into a short rainbow. The technique used by professionals is first to find the orientation of the CCD on the sky by trailing star images across the field. Then well-known doubles are observed to establish the image scale. Amateurs have used cheap webcams and conventional software to make measurements using quite small apertures, though as with visual work the field remains wide open.

There is significantly more interest in double-star observing now, especially in France, Spain, Italy and the USA. A search online will turn up a number of online journals that show what sort of work is being done. It is heavily biased toward CCD imaging of wide pairs which may show common proper motion and would hence be binary systems, albeit of very long period. The more astrophysically interesting binaries with periods of 100 years or less need either a micrometer and reasonable aperture or a webcam-type camera.

This is not the sort of work that people do on a casual basis, so anyone who wants to contribute to knowledge in this way must justify the cost in terms of their own satisfaction. The small number of active amateur double-star observers around the world do find, however, that the work they do is almost certainly not being duplicated by others.

Deep-sky objects

The technique of observing deep-sky objects – that is, clusters, nebulae and galaxies – has changed radically over the past 20 years or so with the introduction of filters designed to cut out light pollution. These are covered in Chapter 6; here I shall deal with observations from the city of the various types of object made using binoculars or a telescope with no filter.

The lessons of Chapter 2 – wait for the clearest nights and aim high – are crucial when you're trying to find deep-sky objects from urban areas. Make sure that there are no lights directly in your line of sight. One neat trick is to block out such lights by hanging a blanket on a clothes line. Some people make screens which they set up when they are observing to shield particularly troublesome lights. The section in Chapter 5 on maximizing contrast within your telescope is also essential reading for the new deep-sky observer. Good baffling inside the telescope can make an enormous difference to your ability to see low-contrast objects.

Don Miles of the Webb Deep-Sky Society – an organization devoted to perpetuating the pioneering deep-sky observing work of the Reverend T. R. Webb in the 19th century – observed for many years with a 200 mm (8-inch) Schmidt-Cassegrain telescope from Lovedean, which despite its idyllic name is actually on the outskirts of the crowded, urbanized area of Portsmouth on the south coast of England. He has the following tips.

Eyepieces in particular get dirty easily, and Don cleans his at least every other observing session. He uses Kodak lens cleaner, which is specially designed for use on coated optics. He claims that he can add up to a magnitude in his light-polluted sky by keeping the optics clean. One advantage of a Schmidt-Cassegrain or a windowed reflector is that the main mirror is not exposed to the atmosphere, so it remains in good condition for longer. The exposed glass, however, is very prone to collecting dew and Don uses a dew cap which projects beyond the end of the instrument a distance of one and a half times the aperture. Not only does this prevent dew from forming, it also keeps stray light off the corrector plate.

Don advises, "Don't think because the sky's bright you don't need to dark adapt. If it's not darker outside than in your lounge, you should pack up anyway!" There are several methods of preserving your dark adaption. One of your most useful accessories can be a piece of black cloth. Use it to cover your head and eyepiece while you are observing, like an old-fashioned photographer. The total darkness keeps your eyes fully dark-adapted and helps you to catch faint detail in the field of view. The long-term improvement in dark adaption may even make you think that the sky has got worse after an hour or so of observing, whereas all that has happened is that your eyes are better adapted.

Alternatively, wear an eyepatch over your non-observing eye. It is always better to observe with both eyes open, to avoid the discomfort of keeping one eye closed. You may even go to the extreme of using one eye for looking at your star charts and the other for observing, using eyepatches to alternate. This is similar to the trick used by World War II fighter pilots who would close one eye when flying into the Sun so as to preserve the vision of their other eye should danger then threaten from a different direction.

Rubber eyeshields on your eyepieces also help to exclude stray light. One very cheap and simple alternative has been suggested by Michael B. Leigh of Lakeside, California. He buys black terry-cloth wristbands which he slips over the end of the eyepiece. They also reduce the risk of image shake, which would happen if you bumped the telescope by getting too close. And if you leave the telescope for a while, he points out, you can fold the band over the eye lens to protect it from dew.

While the country observer can move from one deep-sky object to another according to whim, the town-dweller must plan carefully. Choose objects that will be as near the zenith as possible, and take note of catalog listings that indicate how condensed they are. In city skies, the more condensed objects, even faint ones, are likely to be more visible than extended objects. For example, the magnitude 8.4 galaxy M94 has a condensed nucleus and is much easier to see than the larger galaxy M101, which is given as magnitude 7.5, and supposedly twice as bright. Some catalogs list surface brightness, which is helpful though not infallible. One which does is the *Observing Handbook and Catalogue of Deep-Sky Objects* by Christian B. Luginbuhl and Brian A. Skiff – a constellation-by-constellation guide to some 1,500 objects for medium-size telescopes.

A brief word for the uninitiated about catalog numbers assigned to deep-sky objects. The ones you will encounter most often are those prefixed by "M," "NGC" or, to a lesser extent, "IC." Those with M numbers were originally listed by the French comet-hunter Charles Messier in the 18th century (there were a few later additions to his list), and are now referred to as Messier objects. Messier's aim was simply to note, for his own use, those objects that could be mistaken for comets. He used telescopes which were effectively about as good as a modern 90 mm (3½-inch) refractor. You might imagine that his list would include the brightest objects observable from the latitude of Paris. However, the skies of Paris were then darker than those of modern suburbs, and many Messier objects are now hard to see, except from good locations.

One of the most successful nebula-hunters was William Herschel, who used a telescope with a 470 mm (18.7-inch) metal mirror, with the light-gathering power of a modern 330 mm (13-inch) telescope. The observations from which his catalogue was compiled were made from Slough, to the west of London, and the list was extended by his son John to include southern objects. It was to become the basis of the *New General Catalogue of Nebulae and Clusters of Stars* prepared by

J. L. E. Dreyer of Armagh Observatory. Dreyer's numbers are those we use today, prefixed with NGC. Later additions he included in two *Index Catalogues*, and numbers from them are prefixed with IC. The descriptions and positions have been brought up to date in *NGC 2000.0.*

Clusters

A cluster of stars, whether a loose open cluster or a compact globular cluster, is a fine sight in a dark sky with a good wide-field eyepiece. Sadly, the urban observer has to be content with less spectacular views. Although it's true that increasing the magnification darkens the sky background while leaving star images the same brightness, this also has the effect of reducing the field of view. People pay a lot of money these days for field of view. The new designs of eyepiece that cost and weigh as much as a telephoto lens for a camera can give superb definition over an enormous field of view compared with ordinary Plössls. You almost feel you could climb inside and be there. They are designed in particular to give wide fields with instruments of f/5 or shorter. Other less costly eyepieces can be used on instruments with longer f-ratios. (Eyepieces are described in Chapter 5.)

Any solution that reduces the field of view is not very welcome when the whole point of looking at the object is to gasp with amazement. The city observer must forgo seeing a thousand pinpricks of light against a velvet cushion of sky. But the ultra-wide-field eyepieces do seem to offer the city observer the advantage of their improved contrast. David Cortner of Johnson City, Tennessee, has skies with a naked-eye limiting magnitude of about 4.5 which barely show the Milky Way, even when it is overhead in summer. He has this to say about eyepieces:

"Every time I do a side-by-side comparison, premium eyepieces really do seem to deliver an appreciable boost in contrast, and that's what's most lacking in the city sky. My Tele Vue Naglers and Burgess/TMB Planetary eyepieces make a huge difference in what I can see in many favorite objects. M51 is a vaguely dynamic shape in my old 20 mm Erfle; in my 16 mm Nagler it is a hurricane of soft, milky light, shot through by bright and dark detail. The fact that both my primary instruments are short-focus (f/6 and f/5) may mean that premium eyepieces have more benefits for me than for someone using long-focus optics."

Even with a basic eyepiece, such objects as the Double Cluster in Perseus, the globular cluster M13 in Hercules, and M35 in Gemini are worthwhile targets from urban areas. These clusters contain bright stars which should be easily resolved in a moderate instrument such as a 150 mm (6-inch) reflector. But they are by no means the limit of what you can see. Try looking for the fainter globulars and the smaller open clusters which force you to use a higher power. Even brief listings, such as those in *Norton's 2000.0*, include a selection of non-Messier clusters which can be seen with no difficulty.

With the fainter objects, good seeing may well play a part in turning them from a vague blur into something worth looking at. Many globular clusters should be visible from the city, but you will probably not see as much of them as from a dark site.

Although you may well be able to see faint globular clusters which are well placed, some bright Messier objects can be hard to see. From mainland Britain I have never been able to locate the most southerly Messier object, M7 in Scorpius, even in a comparatively dark sky, because it is always less than 5° above the horizon. I know people who have seen it, but I have never been lucky. James Hilder has seen it from the Galloway Dark Sky Park in Scotland – but needed a 400 mm telescope to do it! Yet from a site farther south, where it rises higher in the sky, it is a very prominent and beautiful cluster, and a spectacular sight in binoculars. When I see it from a location such as Tenerife, in the Canary Islands, I wonder how I could ever miss it from home. This shows the importance of making sure that the object is well placed before you try to observe it.

Star clusters usually contain stars with a range of brightnesses. James Hilder lives in the middle of London, close to the British Museum. From the roof of his apartment he can actually just detect the brightest star in the cluster M6, close to M7, with binoculars and has even seen some stars in M7 using a 250 mm reflector. However, seeing a handful of stars doesn't really do justice to this splendid object.

Many people attempt to see all the Messier objects. This is not a difficult task provided you have the right site, but from a city – particularly in Britain – it is a daunting challenge. The fact is that the list of Messier objects is not a "top 100" of the brightest nebulae and clusters in the sky. It includes many obvious clusters, such as the Pleiades (M45) in Taurus, shown in Figure 4.26 (a), and the Wild Duck (M11) in Scutum, but fails to include the Double Cluster (NGC 869 and 884) in Perseus, one of the showpieces of the sky, and of course most southern-hemisphere clusters such as the splendid NGC 3532 in Carina. Rather than try to catch a glimpse of the fainter Messier objects, which are elusive and unrewarding targets from urban areas, it would be better to look for brighter non-Messier objects which are actually much more interesting.

You will find many beautiful sights in the sky which are not included in any of the main listings. Plenty of groupings of stars visible in low

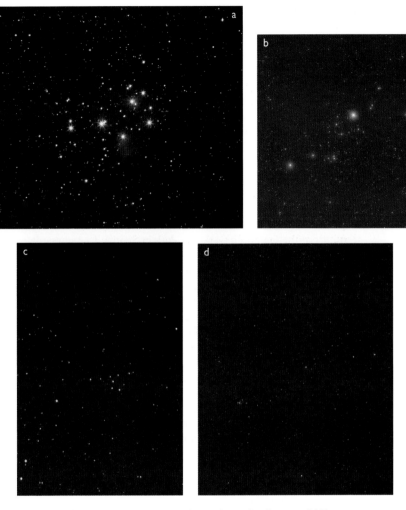

▲ Fig. 4.26 Star clusters can be observed easily from urban locations. (a) The Pleiades (M45), the most prominent of all clusters and a splendid sight in binoculars. (b) One of the unsung clusters in the sky, the Alpha Persei Association. Best seen in binoculars, its stars are more widely spaced than those of the Pleiades. This photo was taken using a diffusing filter to bring out the star colors. (c) The Coathanger, a group of ten stars of sixth and seventh magnitude in a distinctive shape. (d) Kemble's Cascade is a line of stars best seen with a power of about 50×. The cluster NGC 1502 lies at its lower end. Groupings like these last two are not usually labeled as such on star charts, but they are fun to locate and even discover. The Milky Way is a rich area for lines, triangles and circlets of stars that you can make your own.

powers (see Figure 4.26) do not appear in the major catalogs. To observers in earlier days, who did not have to put up with city lights and had much else to look at, they did not seem worth mentioning. One example is the group of stars around Alpha Persei. In binoculars this is as splendid a sight as the Pleiades seen through a telescope. It is actually known as the Alpha Persei Association, and is a genuine cluster of stars about 570 light years away, according to *Burnham's Celestial Handbook*, that excellent source of heavenly goodies, though now outdated in its measurements. Modern sources suggest just over 500 light years. Another well-known cluster is the Coathanger in Vulpecula, which has the obscure catalog number Collinder 399. Even *Burnham's* misses this one, yet it is a delight with a low power. It is, however, shown and mentioned in *Norton's 2000.0*.

Walter Scott Houston, writing in *Sky & Telescope* (November 1991, page 559), gave this description of a line of stars he called "Kemble's Cascade": "In this pallid corner of Camelopardalis, at about right ascension 3h 57m, +63° (2000.0 coordinates), was a celestial waterfall of dozens of 9th- and 10th-magnitude stars. Down it went, tumbling steeply southwestward over 3° before splashing into the open star cluster NGC 1502." To David Cortner in Tennessee, this string of stars is "as much fun to fall along from town as from out in the dark. NGC 1502 at one end of the cascade is easy and rewarding from inside the city."

David Frydman in London gasped when he first saw NGC 457 in Cassiopeia: "It looked just like a praying mantis, with two long lines of stars spreading away from a bright star in the center." Others call this the Owl Cluster, or even the E.T. Cluster from its imagined resemblance to Steven Spielberg's long-armed extra-terrestrial. This cluster is a favorite of mine at public observing sessions organized by the West of London Astronomical Society in Ruislip, Middlesex, where the light pollution makes finding exciting objects a challenge. "Take a look at E.T.," I suggest. It's about the only object in the sky that can actually make people laugh!

Nebulae and galaxies

There is an enormous real difference between a nebula in our own Galaxy and a whole distant galaxy, of course, but to the urban observer both present the same difficulty: they are faint, fuzzy blobs. In general, the nebulae cover a larger area of sky and are more diffuse, while galaxies are more regular in shape – usually elliptical – and have a central condensation. This central region is the brightest and easiest part to see, and indeed with most galaxies it is all that is visible. Seen from the city through a moderate aperture, say 150 mm (6 inches),

▶ *Fig. 4.27 The Andromeda Galaxy, M31, photographed in light-polluted skies then with image processing applied (right) to increase the contrast and correct for the light pollution. This shows how only the nucleus of the galaxy is visible in bright skies – the spiral arms of the galaxy are lost.*

the Andromeda Galaxy (M31) is an elliptical blur about the same angular diameter as the Moon, with another object, M32, some distance away. Compare your view with a textbook photograph and you will see that M32 lies against a background of the spiral arms of M31. What you see as M31 in your binoculars or telescope is in fact just the nucleus. In good skies some observers can see the galaxy extending to 5° or more, instead of a miserable half degree (Figure 4.27). The same applies to many other objects.

Another caution is not to expect too much from telescopic views. The vivid colors of galactic nebulae in photographs in books are mostly real, but the eye is just not sensitive enough to see them properly. Many people can see the Orion Nebula as pale green. I must admit that to me it just looks gray, though I often suspect a reddish tinge to the outer wings when viewing it under good conditions with a moderate aperture. While digital cameras and color CCDs have about the same color sensitivity as the eye, give a long enough exposure and they will show the red. Most photographic films, however, can record the colors very easily, particularly the red color of hydrogen gas, with some types of film being better than others. There is a case for using film instead of digital for some objects (Figure 4.28).

You might, incidentally, imagine that by getting a larger telescope you will be able to see the Orion Nebula in brilliant colors. In fact, even through the biggest telescopes it has the same surface brightness as it does through small ones. No telescope can make an extended object appear any brighter: all it can do is provide enough magnification for you to see it better. You cannot, for example, construct a one-power telescope that will make everything look the same size but brighter. Similarly, even if you were to be able to travel to the Orion Nebula in a spaceship and look at it from close-up, it would still not appear any

▲ *Fig. 4.28 Orion's Belt and Nebula area photographed on the same occasion using a Canon 40D DSLR (left) at ISO 400, and Fuji Provia 400X color slide film. A CLS filter was used in both cases. The Orion Nebula is much redder on the film, though the extent of the nebula is the same in both versions. You can also just see the Flame Nebula to the left of the lower star of Orion's Belt. The film is considerably more grainy, making it more difficult to increase the contrast and bring out the nebulosity.*

brighter. It would be bigger, but you would not be able to see the colors significantly better than if you could make it look the same size through your Earthbound telescope. (Actually, from close by you would have less glass in the way to absorb light, no Earth atmosphere and no interstellar dust, so it would be a little brighter.) If you find this hard to believe, try looking by day at a distant white-painted house or similar object. The white area is just as bright per square degree no matter whether you use the naked eye, low-power binoculars or a high-power telescope. The total amount of light gathered by a large telescope is greater, but it is spread over a larger area. If the house is too remote to register as a finite area on your retina, you simply will not see it at all. The same applies to deep-sky objects.

As with clusters, the Messier list is not a good guide to the visibility of nebulae and galaxies. Compare, for example, two objects visible toward the end of the year: M33 and M76. The galaxy M33 in Triangulum is supposedly fifth magnitude, yet it is very hard to see in light-polluted skies except for a hint of the nucleus, and in such skies it is hardly worth the bother of searching for it. By contrast, the planetary nebula M76 in Perseus is sometimes described as "the faintest Messier object," and is listed as a pair of 12th-magnitude nebulae in *NGC 2000.0*. Yet it is a comparatively easy object to observe because it is very condensed. Writing in *Sky & Telescope* (November 1993, page 104), Walter Scott Houston suggested that it is visually magnitude 10 or brighter. It is less than 5 arc minutes across, compared with about a degree for M33.

The magnitudes given in observing guides are often photographic magnitudes. To the eye, the objects may appear either brighter or fainter, depending on their color. What is more, those magnitudes are a measure of the brightness of the entire object as if it were condensed into a single point of light, instead of being spread over an area of sky. The actual size of the object is far more important. This is particularly true of planetary nebulae, which can vary considerably in size. Apart from anything else, you need to know what sort of eyepiece to use when searching.

Even in perfect skies you need a large telescope to observe the fainter galaxies, and this is doubly true from the city. Aperture is more important than good contrast – a big, cheap Dobsonian is preferable to an expensive refractor. Even so, finding faint nebulae and galaxies from the city can be tricky. It is often fruitless simply to sweep your telescope across the general area of the object in the hope of picking it up. Use a computer mapping program to show the nearby faint stars, and to depict the field of view of your eyepieces, if they can be superimposed on the map.

You can prepare these maps in advance, selecting objects that will be high in the sky, and arranging them in order of right ascension so that you know when each will be at its highest. Ready-printed cards of selected Messier and NGC objects are also available.

Those with computer-controlled telescopes might think that they can simply punch in the number of the object and let the telescope do the work. But even in dark-sky areas, it isn't always as easy as that. Deep-sky objects don't necessarily jump out at you, and the prepared maps can be just as useful. Often, a Go To telescope will not put the chosen object right at the center of the field of view. It is a good idea to "sync" the telescope on a nearby bright star before you search for a particular object, which should improve its finding accuracy.

The fact remains that from a light-polluted environment you just cannot see as much as from a dark site. Even so, you can locate a surprising number of objects. In a five-year period from a site only 6 km (4 miles) from central London, David Frydman observed a total of 600 deep-sky objects, using a 125 mm (5-inch) refractor which he says is the equivalent of a good 150 mm (6-inch) reflector. Of these, 200 he saw only once, but the others he observed at least twice.

5 • CHOOSE YOUR WEAPONS

You can be a successful amateur astronomer without owning any form of optical instrument. There is satisfaction to be gained just from naked-eye observing, even with a light-polluted sky. You can strain to catch a glimpse of the Double Cluster, or the Orion Nebula. But after a while the excitement of doing this palls, and your thoughts turn to buying a telescope. For the suburban astronomer, as opposed to one living in the country, there is more to consider and perhaps less freedom of choice when it comes to buying equipment. To get the best results, the apparatus must suit the observing conditions.

To most people it seems obvious that the one thing an amateur astronomer needs is a telescope. Yet a few leading observers rarely use telescopes – instead, they prefer binoculars. It is now standard advice to beginners that the ideal starting instrument is not a telescope, but binoculars.

Binoculars

Binoculars have many advantages over a telescope. They are comparatively cheap because they are mass produced, and are much better value for money than a very small telescope. They do not need an expensive mount – your arms are good enough for most purposes. They reveal a wide range of objects in the sky that are quite invisible to the naked eye. And you can of course use them in the daytime for a wide variety of other purposes.

I'll give some tips about buying binoculars, since few books go into the matter in any detail. Binoculars are rated by the size of their main lenses – their *objectives* – and their magnification. A typical specification is 10 × 50, pronounced "10 by 50." These have a magnification of 10× (pronounced "10 times") and objectives 50 mm (2 inches) across. It is the size of the objectives that determines how bulky a pair of binoculars really is. A pair of 7 × 50s is virtually identical to a pair of 10 × 50s, which have a higher magnification.

The standard size of binoculars for general-purpose observing is usually said to be either 7 × 50 or 10 × 50. Of the two, urban observers will probably choose 10 × 50, since the extra magnification will help to give a slightly darker background. The 7 × 50s come into their own in really dark skies, and notably for younger observers whose pupils can open wide enough to admit the additional light. But there is a place for other sizes, such as 8 × 30, which are noticeably less heavy. This is important if you observe for any length of time without support, for it is better to have a pair you can use with comfort than a pair you frequently have to put down to rest your arms. I would advise against the more ambitious sizes, such as 20×80, until you are sure you need them. However, 15×70 binoculars are now particularly good value and the extra magnification is excellent for revealing the smaller objects such as galaxies. But I wouldn't recommend them as your only binoculars.

You need not pay a great deal for binoculars, but you should certainly not go straight for the cheapest. Unlike telescopes, you can often test them before you buy if you can find a stockist with a good range. Try a number of different models, including the more expensive ones, so you know what's good and what's bad. There are some very high-priced binoculars on the market aimed particularly at the lucrative birdwatching market. Their prices are pushed up by features such as coatings on all the glass surfaces and additional lightness and ruggedness, which are not critical in astronomical use. Astronomers rarely observe in the rain, so good waterproofing is not important! Birdwatchers have other requirements, such as being able to see a bird against a bright sky, which is where the extra coatings help. Very expensive birdwatching binoculars may not actually perform as well on the night sky as ordinary 10×50s.

Unlike telescopes, binoculars can be tested reasonably well during the day (though stars are always the best test for astronomical purposes). To evaluate a pair of binoculars, choose a test object such as a TV antenna that stands out against a bright sky, but away from the direction of the Sun. Look for signs of false color on either side of the object. Slight color fringing, usually apple green and magenta (pinkish), is inevitable, but it should be noticeable only if you look hard for it. Turn slightly so that the object moves to the edge of the field of view. The object should remain sharply in focus.

Cheap binoculars usually have a limited field of view – that is, the image you see is surrounded by a large area of black, like looking into a tunnel, rather than giving you a wide perspective as in the more expensive models. But a wide field of view is useless if the definition is not good all the way across. Avoid wide-field eyepieces that may give a curved plane of focus. Check for this by examining a flat wall at right angles to you – it should be perfectly in focus from edge to edge. Also make sure that the magnification is even across the field of view by panning across it – avoid binoculars with barrel distortion, which makes straight lines near the edges of the field bow outward, and will make star patterns near the edges of the field of view seem to bulge toward you as you sweep across the sky.

Check for vignetting (uneven illumination of the field of view) by looking through the binoculars the wrong way. You should see the circular outline of the eyepieces even when looking through from the very edge of the objectives – the outline should not be cut off by the edges of the prisms inside. Very lightweight but cheap binoculars may not be rugged enough to prevent the prisms from becoming mis-aligned by an accidental knock, and you could be faced with a repair bill for realigning them that is more than they cost.

If you wear spectacles for short or long sight, make sure you can focus on infinity using the binoculars with your spectacles off. But if your spectacles are to correct astigmatism, you will have to keep them on while observing and you will need binoculars with rubber eyecups that can be turned back to give a soft surface against which you can press your spectacles.

There is also the question of *exit pupil* to be considered. The exit pupil is the diameter of the beam of light that comes out of the eye-piece and into your eye. For a pair of binoculars it is easily calculated – it is the size of the objectives divided by the magnification. So the exit pupil for a pair of 10 × 50s is 5 mm, and for a pair of 7 × 50s it is just over 7 mm. You can see the difference if you hold the binoculars away from you so that you can see the disks of light in the eyepieces.

When your eyes are dark adapted, the pupils open up from their daytime diameter of about 3 mm to about 7 mm. This figure is an average for the population at large, and values for individuals vary from about 8.5 mm down to 4 mm, even among young, healthy people. There is a definite general decline with age, and the average for 80-year-olds is about 3 mm. There is no point in using 7 × 50 binoculars, which have an exit pupil of 7 mm, in order to see a brighter image if your pupils only open to 5 mm. So it could be worth you checking your pupil size before you buy. The simplest way is to get someone to measure them with a ruler held up close to your eye in dim light.

Even experienced telescope owners still use their binoculars, so you should expect to get good use out of them. They are fine for locating such objects as comets and novae, and simply for the pleasure of gazing at the Milky Way. Binoculars can reveal a good range of interesting objects, even from the city. But most amateur astronomers find that the low-power views through binoculars are not all they want, and inevitably they begin to think of getting a telescope.

Which telescope?

Astronomers love to talk about telescopes and to compare one with another. And, as in any specialist discussion, there is as much nonsense talked as there is truth. When it comes down to it, the "best" telescope is as impossible to define as the "best" car. In astronomy, as with cars (and other things), it's not what you've got, it's what you do with it that counts.

Of the three basic types of telescope – the refractor, the reflector, and the catadioptric – each has its own peculiarities and advantages. Even so, whatever type you have and wherever you live, you can still accomplish a great deal. But if you want to get the most out of urban observing, it helps if you use the most suitable instrument. Because so many people ask for guidance on buying a telescope, I'll go into some detail about the basic types and what they can do.

Refracting telescopes

The refracting telescope is the popular concept of what a telescope should look like, with a big lens – the objective – at one end pointing skyward, and a small lens – the eyepiece – at the bottom of the tube, through which you observe. The objective focuses light from distant objects to form an image, and the eyepiece gives a magnified view of the image. To magnify it more,

you use a more powerful eyepiece. The size of a telescope is usually indicated by the diameter of the objective, as the magnification it gives can be changed by using a different eyepiece. A magnification of 100 times, say, is often called a power of 100 or 100 power, and is often abbreviated to 100×, or ×100. The higher the magnification, the higher the power you are using. Figure 5.1 shows a typical modern refractor.

However, there is one drawback that prevents most refractors from being the ideal telescope: *false color*. A single lens, such as a magnifying glass, splits light into the colors of the rainbow as it bends or refracts it toward a focus. This is not very

▶ *Fig. 5.1 This Sky-Watcher 120 mm (4.7-inch) refractor is on an EQ5 Go To equatorial mount with a built-in pole-finding system.*

noticeable in a magnifying glass, which magnifies only a few times, but in magnifying the image many times, as in a telescope, the false color becomes objectionable, and some means has to be devised to reduce it. The quality of a refractor is to a large measure determined by how, and how successfully, this is done.

Very basic telescopes sold by a wide range of outlets from mail-order houses to department stores are what is called *non-achromatic*: they have only a single-element objective lens, which itself does nothing to correct for the false color it generates. The aperture is generally *stopped down* – that is, there is a disk with a hole in it some distance down the tube, which considerably restricts the instrument's effective diameter. The purpose of this stop is to reduce the inevitable false color in the image. But it also makes the image so dim that most of these telescopes fail to show any but the brightest stars. The result – disillusionment with astronomy. I mention such telescopes only because they are widely available and are often bought for children. They are not necessarily very cheap – with the packaging and unnecessary extras they may cost as much as a low-end digital camera. And even some very reputable department stores may sell them in ignorance of their true worth.

The next stage in improving false color is to make the objective lens from two glass elements – a *doublet* – arranged so that the color dispersion of one counteracts that of the other. Such a lens combination is called an *achromatic lens*, or simply an *achromat*. This is the system used in a wide range of refractors costing anything from the price of a compact digital camera to that of a large TV. At the cheap end the quality often leaves something to be desired: either the false color is not properly corrected, or there are other image imperfections which make the telescope suitable only for very low magnifications, below 20×.

Even well-made basic refractors suffer from some false color, though modern instruments do perform much better than many of those made in the last century. With the most readily available types of optical-quality glass it is possible to correct for only some of the false color, and in my experience a typical refractor gives objects a bluish halo. The better the telescope, the less objectionable the halo and the higher the magnifications you can use. A 100 mm (4-inch) refractor of this type will cost about the same as a medium-sized TV, and should give crisp views up to a magnification of 150×.

The ultimate in color correction in refractors comes from using advanced materials in place of glass for one or more of the elements of the objective lens. These materials have different refractive properties from ordinary glass. Originally, artifically grown crystals of calcium fluorite were used, but these were delicate and have now given way to other materials with extra-low dispersion, known as ED glass. The more expensive of these telescopes can give virtually color-free images, and have price tags similar to those of small cars. Objective lenses that give excellent color correction are called *apochromatic* or simply *apo* lenses, though these terms tend to be somewhat misused and there are differing degrees of excellence. Just using an ED glass such as FPL-53 does not guarantee a color-free refractor, as I've discovered to my cost!

One additional advantage of refractors is that they can provide a wide field of view free from aberrations such as coma, which bedevils Newtonians in particular. This is particularly useful for photographers, especially when using DSLRs, and small refractors in particular have seen a resurgence of interest. Whereas at one time refractors of 60 mm to 80 mm aperture were regarded as suitable only for beginners or for traveling with, today they are useful workhorses for photographing a wide range of deep-sky objects and comets.

Some refractors are designed with very flat and wide fields of view and are known as astrographs, though the term is becoming used to describe any instrument, refractor or reflector, that is designed primarily for imaging.

Reflecting telescopes
In a reflecting telescope the image is provided by a concave *primary mirror* rather than a lens. The great thing about mirrors is that they reflect all colors equally, so there is no false color in the image. However, the image is formed right in the middle of the incoming beam of light, so a smaller *secondary mirror* is placed so as to reflect the light to the side of the tube, where the image is viewed through an eyepiece. This basic design of reflector is called the *Newtonian* (see Figure 1.3 on page 10).

This arrangement has important consequences. First, you view sideways on to the object you are observing, which does not affect your view in any way, but makes finding objects a little more tricky than with a refractor. In order to aim it you have to shift your observing position to squint along the tube or look through the finder telescope, which is either a low-power refractor or a red-dot finder. Second, the flat secondary mirror both blocks a little light from the main mirror and introduces the effects of diffraction into the image. *Diffraction* is a slight bending of light as it passes an obstacle, in this case the secondary mirror and its support. Because a small proportion of the incoming light is diffracted, and does not go where it is wanted, the contrast of fine details is reduced a little, which affects planetary observing in particular.

Third, a little light is lost with each reflection. While this does not matter too much when the mirrors are freshly coated, after a while the percentage of light reflected will decline from about 88% when new to, say, 82% as the coating's reflectivity deteriorates over a period of a few years. Since there are two mirrors, the amount of light transmitted to the eyepiece will be reduced from 88% of 88%, which is about 77%, to 82% of 82%, which is about 67%. Over time, each mirror will also tend to collect a layer of dust specks, further reducing the contrast and light transmission. More reflective coatings are available, but the dust problem still applies. Cleaning a mirror is no simple task – the mirrors must be removed from the telescope first, and the surface coatings (usually aluminum) must be treated with care. By contrast, the objective lens of a refractor absorbs only about 2% of the incoming light, while reflection from the glass surfaces can be reduced to a similar percentage by the use of anti-reflection coatings. The total light transmission of a good refractor is in excess of 90%, and, unlike that of a reflector, it remains close to that figure. Any dirt on the outside surface of the objective is comparatively easy to access and to clean off.

The tubes of reflectors are usually open at the top end, providing access not only for dust and other airborne particles, but also for stray light. Air currents are far more likely to circulate in the open reflector tube than in the sealed refractor, and the thick mirror of a reflector can take longer than a lens to reach night-time temperature when it is carried from indoors out into the night air, resulting in disturbed images. The mirrors of reflectors can easily go out of alignment, making them less straightforward to use and maintain than refractors. And reflectors suffer from the aberration of coma which gets increasingly worse toward the edge of the field of view, though coma correctors which fit at the eyepiece end are available.

All these drawbacks would seem to make reflectors a poor choice, but they have two saving graces. First, they are much cheaper to make in larger sizes than are refractors. Second, the lack of false color means that a reflector can give a superior image to many refractors. For many amateurs, therefore, the reflector is the choice. The same amount of money that will buy you a rather average 100 mm (4-inch) refractor will get you a 150 mm (6-inch) reflector that will show finer detail on planets and fainter deep-sky objects. The price differential becomes much greater as you go to larger sizes, and refractors larger than 175 mm (7-inch) aperture are both very expensive and rare.

Catadioptric telescopes

Both refractors and reflectors are rather long for their diameter. This feature, broadly speaking, is called the *focal ratio* or *f-ratio*. Properly defined, it is the ratio of the focal length of the objective lens or mirror to its diameter, the resulting number being known as the *f*-number. (The focal length is the distance from the lens or mirror to the focus, the point where the image is formed.) So an *f*/15 telescope is very long compared with its diameter, and an *f*/4 telescope is short and squat. The squatter the telescope, the easier it is to handle and to mount. But there is a price to be paid for this convenience: such telescopes are more expensive, and perform differently from longer *f*-ratio instruments. In general, the shorter the telescope, the lower the magnification it will give with a particular eyepiece. But the catadioptric design, which combines reflecting and refracting optics, overcomes this problem to give a short-focus but powerful telescope that is easy to mount and to carry around.

Probably the most common design of catadioptric is the *Schmidt-Cassegrain telescope*, or SCT, as popularized by manufacturers Celestron and Meade (see Figure 5.2). A telescope with a standard 200 mm (8-inch) mirror and a focal ratio of 10 would normally need a tube 1.5 to 1.75 meters (say, 5 to 6 feet) long. An SCT has a glass corrector plate – a thin lens – at the top end of its tube, closing it in. The secondary mirror reflects incoming light back down the tube (not to the side, as in the Newtonian) and through a hole in the main mirror, behind which is located the eyepiece. You observe in the same direction as the object, as in a refractor, but the tube is only some 600 mm (24 inches) long.

The SCT design packs a lot of telescope into a small package, and because of its compactness the mounting can be a lot lighter than for a refractor or reflector of similar

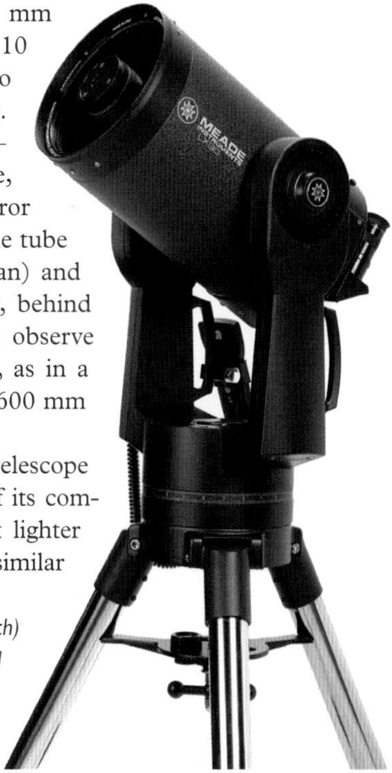

▶ Fig. 5.2 *A Meade LX-200 250 mm (10-inch) Schmidt-Cassegrain. The telescope is controlled by a purpose-built computer which can locate and follow objects, even when used as an altazimuth instrument, as shown here.*

aperture. A whole industry has grown up around making accessories for these popular telescopes.

An SCT works out at over twice the price of a Newtonian reflector of the same aperture, but many amateur astronomers find that the convenience of the design is worth it. Optically, though, there are shortcomings. The worst is that the secondary mirror is larger than it would be in a Newtonian, so the effects of diffraction start to become more evident. Contrast of fine detail is worse than in a Newtonian. In the SCT's favor, however, is that the closed tube should keep the main mirror free from dust, make the instrument quicker to settle down when taken outside, and reduce air currents in the tube.

The other main catadioptric design is the Maksutov, which has a curved corrector plate and tends to have long focal ratios well suited to planetary observing. These can be made either as Maksutov-Cassegrains, with the eyepiece behind the mirror, like SCTs, or Maksutov-Newtonians, with the eyepiece at the side like a Newtonian.

Other essentials

A good telescope has many essential components besides its optics, and a shortcoming in any one of them can destroy its usefulness. The experienced observer knows that a telescope's mounting is vital. Many cheap telescopes are let down as much by an insubstantial, wobbly mounting as by other failings. Some amateurs maintain that the eyepiece is just about the most important part of the telescope, and indeed it is true that a good telescope can be ruined by a bad eyepiece. In each case, though, it is rather like claiming that one of the legs of a three-legged stool is more important than the other two.

Mountings

Ideally, what you want from the telescope's mounting is that you should be able to swing the instrument so as to bring it to bear on your target, and find that target within seconds. The drive, a mechanism that slowly turns the telescope to compensate for the apparent rotation of the sky, should be sufficiently accurate for you to follow the target for as long as you wish. You should be able to focus easily without the telescope taking minutes to stop shuddering, and a gentle breeze should be able to waft past without the image breaking up into dancing patterns.

There is always a tendency to keep costs to a minimum, and a telescope that is manufactured to a price cannot be expected to be rock steady. I won't discuss mountings in any great depth because the choice has little to do with whether you observe from the city or from the country. The main thing is to make sure that the mounting is well enough engineered for your observing not to be restricted by it.

Your basic choice is between the simple *altazimuth* mounting, which allows the telescope to move up and down (in altitude) and from side to side (in azimuth), and the *equatorial mounting*. In this, one axis, the polar axis, is aligned on the celestial pole, so that motion of the telescope about the other axis is parallel to the celestial equator, like the apparent movement of the sky, enabling objects to be tracked more easily.

The advantage of the altazimuth mounting is that it is cheap, so you can buy a steadier mounting for your money. From an engineering point of view, the stresses are easily dealt with, and these days the very largest professional telescopes have altazimuth mountings. The popular *Dobsonian* telescope, shown in Figure 5.3, is essentially a Newtonian reflector on a redesigned altazimuth mount which is easy to make. Its main advantage is its cheapness, and it is an excellent way of mounting a large instrument at low cost. As a functioning telescope, however, it is not particularly portable and is best suited to fairly low magnifications. Having said that, I would rather have a good 150 mm (6-inch) Dobsonian with a smooth and steady mount than an average 100 mm (4-inch) refractor on a spindly tripod and undersized undriven equatorial mount, even for planetary observing. In fact, undriven equatorial mounts (that is, without electric motors) are of questionable value, unless the hand-operated slow motions are exceptionally good, which they usually aren't. If you want to carry out long-exposure astrophotography, however, a driven equatorial mount is pretty essential.

Go To mountings have come of age, and are widely available at reasonable cost on even small instruments. For the urban astronomer they are a boon, as they make it much easier

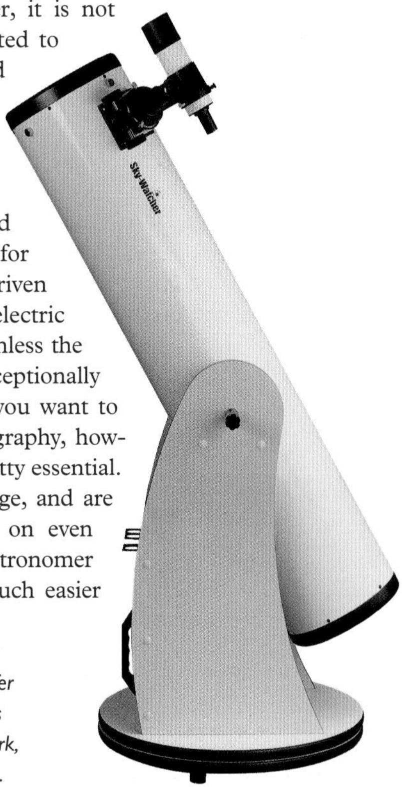

▶ *Fig. 5.3 Dobsonian reflectors such as this 200 mm (8-inch) Sky-Watcher instrument offer low-cost observing. The altazimuth mounting is more suited to deep-sky than to planetary work, and photography is limited to short exposures.*

to find objects in skies where few stars are visible. The usual method of finding faint objects is called star-hopping, which involves starting from an easily found object and then navigating using fainter stars. But where fewer stars are available, this gets difficult and Go To should at least get you to the right area, even if not straight to the object.

Eyepieces

A telescope's eyepiece is what provides the magnification. To vary the magnification (often called the power), you use a different eyepiece. Like some other things in astronomy (such as the magnitude scale), the figures work in reverse from what you might expect. Long-focal-length eyepieces give the lowest magnification.

Most telescopes come with a standard eyepiece of around 25 mm, which gives a different magnification with different telescopes, depending on the focal length. Divide the focal length of the telescope by that of the eyepiece to get the magnification. So a 25 mm eyepiece on a telescope of 1,000 mm focal length gives a power of 40× (pronounced "40 times").

The majority of eyepieces supplied with good telescopes these days are Plössls, though Kellners are found on the cheaper reflectors, and Huygenians and Ramsdens are also used on small refractors. (For a simple outline of the various eyepieces, refer to a manual of amateur astronomy such as *Norton's 2000.0.*) A good rule of thumb is that if your telescope is f/8 or longer, you can get away with simpler eyepieces than if it is f/6 or shorter. A simple and cheap eyepiece design such as the Kellner is quite adequate for much general viewing, particularly at the lower powers and with fairly long f-ratios. For the higher powers, orthoscopics are also perfectly suitable. An Erfle eyepiece offers a fairly wide field, though the definition at the edge is not very good. Plössls can give better definition over a wider field than orthoscopic eyepieces which used to be the standard, but the name is now applied to a wide range of designs.

◀ *Fig. 5.4 A 20 mm Nagler eyepiece (right) and a conventional 20 mm Erfle eyepiece. The Nagler, with its 50 mm (2-inch) barrel, gives an apparent field of view of 82°, while the Erfle, in a 31.7 mm (1¼-inch) fitting, has an apparent field of view of 60°. A similar-sized 25 mm orthoscopic eyepiece, by comparison, has a field of view of only 36°.*

The standard eyepiece barrel size is now 31.7 mm (1¼-inch). But many telescopes now accept 50 mm (2-inch) eyepiece barrels, with adapters to take standard eyepieces. The larger barrel is really needed only for eyepieces of long focal length or wide field, but also makes it easier to attach cameras without risk of vignetting the field of view (that is, causing a drop-off of light at the edge of the field).

Ultra-wide-field eyepiece designs give an apparent field of view of 80° or more, and some observers will use nothing else, even for planetary viewing where the wide field is not important.

Most eyepiece sets include a Barlow lens. This is a negative lens in an eyepiece barrel. It fits between the telescope and the eyepiece, and it has the effect of increasing the focal length of the telescope by a set factor. So with a 2× Barlow, a 500 mm focal length becomes 1,000 mm. This effectively doubles the magnification of any eyepiece, which gives you twice as many magnifications as you had before. When you are choosing extra eyepieces, take this into account. So if you start with a 26 mm standard eyepiece, get a 2× Barlow and, say, an 18 mm eyepiece. That will give you four different magnifications for the price of three: 26 mm, 18 mm, 13 mm and 9 mm. A Barlow is also very useful for increasing the effective focal length of a telescope for photography, as covered in Chapter 6.

Finders

A finder is a device that helps you to find your target, and was traditionally a small telescope with a wide field of view, attached to the tube of the main one, though red-dot finders are now ubiquitous on the cheaper instruments. But a good optical finder is essential for urban observers. Red-dot finders, which have a plain window through which you can see a red dot on the sky, are fine for rough aiming, but in my experience they are useless for actually finding anything other than bright objects.

Every urban telescope needs an optical finder, ideally 8 × 50 or similar. This will show you stars fainter than you can see by eye alone, which is needed if you are to find anything at all. Only Go To telescopes of proven accuracy will find objects without a good optical finder, and even then many red-dot finders are so poorly made that the dot shifts as you move your eye, making it hard to know when you are lined up on an alignment star. A good optical finder really is worth having, and what's more, the battery never runs down! However, a red-dot finder, or the more elaborate type such as the Telrad, which gives you a bull's-eye apparently projected on to the sky (Figure 5.5) can be handy for aiming the telescope in the first place.

If there is one piece of essential advice for newcomers to astronomy, it is this: make sure your finder is properly aligned before you do any

◀ *Fig. 5.5 The view through a Telrad finder, which projects rings into the line of sight over the telescope tube. This helps you to align the telescope on faint objects whose position you know relative to stars that are visible. The rings appear to be at infinity and can be viewed either by eye or using binoculars, which is particularly useful in urban skies.*

observing, even if you have a Go To telescope. Find some object by day in the main telescope, then adjust the finder to point to it. If it's so badly made that you can't adjust it properly, get a better one. And the second piece of advice is always to use the lowest magnification eyepiece first – usually the standard eyepiece of about 25 mm focal length – and then move on to the higher magnifications once you've found your target.

Imaging requirements

The term *imaging* is widespread, but basically it just means photography, which now practically always uses an electronic chip rather than film. There are several different types of camera that you can use, the main ones being consumer digital cameras, usually DSLRs; webcam-type cameras for planetary, lunar and solar imaging; and specialist CCD cameras. Each has its own advantages. For planetary imaging, any driven steady mount will do, but for deep-sky and other faint-object photography, you usually need an equatorial mount whose polar axis is well aligned on the celestial pole (making it parallel to the Earth's axis), a smooth and accurate drive system, and a means of checking for and correcting the drive errors that inevitably creep in. For the latter, autoguiders are available which do the job for you.

Aligning the polar axis on the pole is made easy on the many commercial mountings that have pole-finding systems, as shown in Figure 5.6. For convenience, however, it is best to have the mount permanently fixed in place. That way, you do not lose valuable observing time in setting up. Too often I have either finished setting up just as the clouds have rolled over, or taken a chance on a quick alignment that subsequently proved to be slightly out. The longer the focal length of your telescope and the longer the exposures you plan to do, the more accurate your alignment needs to be. For exposures of a few seconds, either on the planets using film or on deep-sky objects using a CCD, rough and ready alignment may be good enough.

▶ *Fig. 5.6 A pole-finder consists of a small telescope inside the polar axis of an equatorial mount. After turning the axis to the correct angle for the date and time when you are observing, you adjust the mount so that Polaris is within an illuminated circle in the pole-finder field of view. In the southern hemisphere there is no pole star, so you have to use much fainter stars instead.*

Making your own

Amateur astronomers have enjoyed making their own telescopes ever since William Herschel decided he could not afford the telescopes on sale in 18th-century Britain, and found he could make a better one himself. These days there is little economic advantage in building your own telescope, though some people do it because it can be very rewarding. It is more common, however, to create your own setup using individual components, with the telescope (often referred to as the OTA, for optical tube assembly) from one manufacturer, the mount from another, and almost invariably the camera and associated connectors and brackets from yet another.

The urban astronomer's choice

The one type of telescope that is not particularly good for visual observing from towns and suburbs is the very short-focus reflector. Really short *f*-ratios, such as *f*/4, are best suited to observing deep-sky objects at fairly low powers, with the potential for wide fields of view. In a bright sky, much of this potential is wasted. This is not to say that a short-focus reflector is useless – it is just not the most appropriate instrument for the visual urban astronomer. However, for photography or CCD work, its additional speed and wide field are a definite advantage.

Some people would advise, with justification, that the best telescope for urban use is a perfectly corrected refractor of around 150–180 mm (6–7 inches) aperture. Such a telescope is ideal for observing those objects least affected by light pollution, namely the planets. A good refractor gives the highest possible contrast, which is important if you want to see fine planetary detail and also if you wish to search for small deep-sky objects. The modern designs allow for quite low *f*-numbers (around *f*/7 or *f*/8), which means that you can use fairly wide-field eyepieces. The shorter tubes of these telescopes makes them not so massive as to require a permanent observatory.

But there are usually other considerations, notably cost and space. For most people, these matter a great deal. Not only is a top-quality 150 mm (6-inch) refractor expensive, but it needs to be mounted quite high so that you can view through the eyepiece without crouching, even with a star diagonal. The pedestal will probably need to be as high as you are. Unless you are either dedicated or fortunate enough to own one, you will have to choose between less expensive and possibly more portable instruments. What about standard Newtonian reflectors?

I shall nail my colors to the mast here and say that my own preference is for the reflector. While refractors are traditionally noted for their better contrast and image stability, I have always thought that a reflector is a much better choice (unless money is no object). You get more aperture for your money with a reflector, and it is usually easier to observe with. As for the supposed lack of performance, in my experience what counts is focal ratio. The higher the focal ratio, the smaller the secondary mirror compared with the main mirror, and the smaller the central obstruction. There will also be fewer traces of coma, which is the tendency for star images toward the edge of the field of view to appear as small streaks instead of points of light.

In practice, a well-made reflector will deliver superb planetary images at a fraction of the cost of a refractor of similar aperture. Although, size for size, the refractor has the edge, the difference is much less than it was once held to be. In a detailed test report published in *Sky & Telescope* (March 1992, pages 253–7), Douglas George concluded that a good-quality 150 mm (6-inch) reflector approached the performance of a 175 mm (7-inch) refractor. And since the reflector costs much less, it wins my vote. The key is in the quality of both optics and manufacture. Do not buy the cheapest reflector available and expect top performance. If you are prepared to spend more, you should get excellent performance at a fraction of the cost of an apo refractor. If you have the choice, aim for an *f*/8 instrument rather than *f*/5 or *f*/6.

There is a further option. The main advantage of reflectors over refractors is that they can be made in larger sizes without the cost rising steeply. You can buy commercially made Dobsonians of enormous aperture for comparatively little money: telescopes of 430 and 500 mm (17 and 20 inches) aperture are readily available at about twice the price of a basic SCT. The optics may not be perfect, but the aim is to bring in as much light from deep-sky objects as possible – the so-called "light bucket" approach. While this may seem to be pointless in an urban environment, there are advantages. After all, if the objects you are searching for are faint anyway, the more light you can pull in, the better. So the very large reflector also has its place in the urban astronomer's armory.

David Cortner, who observes from a light-polluted part of Tennessee in the US, owns both a 400 mm (16-inch) *f*/5 reflector and a 125 mm (5-inch) *f*/6 Astro-Physics fluorite refractor. He says:

"As a card-carrying member of the 'A-P [refractor manufacturer Astro-Physics] League,' let me say that the best views I get from my light-polluted backyard come from my light bucket, not from my refractor. The only way you'll see galaxies looking like galaxies is with aperture, and aperture is not as crippled by suburban light pollution as I expected it to be. My 16-inch will get 15 to 16th mag. easily in the country, but from the city it still pulls in 14.0 to 14.5, even when the sky is pale and unexciting. Vaguely quantitatively: the naked-eye light-pollution tax here is maybe three magnitudes, but the telescopic penalty is nowhere near as severe, and is less so the more aperture I bring to bear."

What about the SCT? There is no doubt that the contrast is not as good as in a refractor or long-focus reflector, despite the many testimonials to the contrary that the manufacturers use in their advertising. However, the real advantage of the SCT lies in its convenience and the availability of accessories. You will get better performance with a well-made and well-maintained SCT than with a reflector with a dirty mirror. Computer-controlled SCTs are very popular. With very little initial alignment, you simply center a known star in the field of view, repeat with another across the sky, and then request the computer handset to locate any object in its vast memory bank.

A telescope for the city and suburbs

Whatever telescope you own, there are ways of optimizing your telescope for city life, in particular by making sure that it gives maximum contrast. The name of the game is signal-to-noise ratio. The signal is the feeble light from the object, and the noise is the vast flood of light from the sky in general, from streetlights, from security lights, and from your neighbors' homes. Your task is to maximize the signal and minimize the noise. Even observations of bright objects, such as planets, will benefit.

Keep it clean

Getting the maximum signal means keeping your optics as free from dust as you can. Use a cover that keeps all dust out when the telescope is not in use, and line the cover with blotting paper or some other absorbent material that will help soak up any dew that may be on the optics after a night's observing. There will need to be an air hole in the cover so that the optics can dry out, or you will be trapping the moisture. A good material for making sure that the cover fits well is self-adhesive velour, used for the bases of table lamps, and available in rolls in hardware stores, or the soft part of self-adhesive Velcro tape.

Inevitably, dust will collect on the surface of mirrors and lenses. And it doesn't just lie there – it gets stuck on. Do not try to blow it off, for you could just end up spitting on the glass. Compressed air cans are as bad or worse, as they often leave residues. Photographers use rubber-bulb air blowers for cleaning lenses, ideally with nozzles that concentrate the air stream. There are also lens-cleaning pens that use a thin layer of carbon, claimed not to harm the lens coating, to soak up grease.

A little dust can be lived with. If you are happy with the results you are getting, it may be better to leave the dust alone than to try cleaning it off. Eventually, however, the day arrives when it has to go. If you have a refractor or an SCT, use a proprietary lens-cleaning fluid and a clean handkerchief to remove it. There should never be any need to remove the objective lens from a refractor, so that is all the maintenance to the optics you will need to do. A mirror, however, has a more delicate coating, and you will have to remove it from its cell, which usually means removing the cell from the telescope first. You should clean your mirror at least once a year if there is an obvious buildup of dust. Cleaning actually prolongs the life of the surface. Dust absorbs moisture and acts as a site where contaminants can gather and eventually corrode the mirror. This is why tiny pinholes eventually appear on a mirror, even if you never touch it.

Immerse the mirror in warm water in which a small amount of wetting agent has been thoroughly dissolved. Photographic wetting agent is ideal, or you can use a non-biological detergent such as Dreft. Do not use dishwashing liquid or rinse aid as this often contains other ingredients, such as lanolin, which make it gentle on the hands but can leave a greasy residue that may well be worse than the dust you had in the first place. In hard-water areas, use distilled or deionized water as supplied for car batteries or steam irons.

Even after a good soaking the dust may be reluctant to shift, so you will have to swab the glass gently with fresh cotton wool, changing it frequently to make sure that you do not drag sharp particles across the surface. Finally, choose a dust-free place for the mirror to dry. Support it resting on its edge on a wad of clean absorbent paper, so that the water drains off quickly and is soaked up.

Now you will have to realign the optics. You have to make sure that when you look through the eyepiece hole, the outlines of both the flat and the main mirror are concentric, and that the reflection of the flat in the main mirror is dead central. It is a good idea to put a small spot at the center of the main mirror to help you align the optics. It will always be in the shadow of the secondary, so there is no need to remove it afterward. Alternatively, make a cross of string at the top and bottom of the tube to help you get everything in line. Realignment is a detailed procedure, and is dealt with at http://www. stargazing.org.uk/collimate.html and many other online references.

Down with noise

Assuming that your optics are clean, you now need to cut out unwanted stray light – the "noise." For all telescopes, a good long dew shield or tube extension is essential for urban observing. With refractors and SCTs, it shelters the objective or corrector plate and helps prevent the formation of dew, and with both refractors and reflectors it helps to prevent stray light from entering the tube. This is particularly important with Newtonian reflectors, as you will see if you look into the eyepiece tube at night with no eyepiece present. As well as the rather small area of the night sky in the main mirror, you may see a large expanse of the opposite side of the telescope tube, the support for the secondary mirror and maybe some of the focuser itself by means of ambient light. Around the reflection of the main mirror there may be a visible part of its cell. Ideally, you should not be able to see these parts of the telescope at all. They are just contributing stray light which reduces the contrast of your image.

By fitting a dew shield or tube extender, you are restricting the light that enters the telescope more closely to that from the area around the object itself. Make sure the dew shield does not encroach on the field of view of the telescope. If it were the same diameter as the telescope tube and were too long, it would reduce the effective aperture. A dew shield and other ways of fine-tuning your telescope's performance are shown schematically in Figure 5.7.

Inevitably some stray light does get in, and steps should be taken to prevent this light from reaching the mirror or eyepiece. The inside of the tube can be painted matt black, but make sure that the paint really is matt. A smooth surface such as PVC is really quite difficult to dull down, and the slightest shininess in the paint will create reflections. If you can manage it, it is a good idea to fit internal ring-shaped baffles inside the main tube and focuser tube to cut down any further reflections. These are often fitted to refractors, but less often to reflectors. Do not, however, use the self-adhesive velour mentioned above to line the tube. This will provide an insulated layer around the inside of the tube that will retain heat, giving you trouble from tube currents. Baffling the end of the focuser tube, so that the eyepiece sees only the light from the main mirror, can cut out stray light that reduces

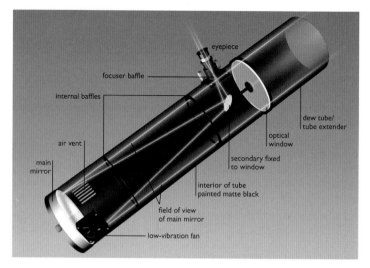

▲ Fig. 5.7 Ways of improving the contrast and performance of a reflecting telescope. Extending the tube and painting the inside of the tube black are the simplest and most effective measures. Baffles cut down the amount of stray light that reaches the main mirror. The fan and the optical window are alternatives: a fan is not needed if the tube is closed by a window, with the tube also blocked at the lower end. A fan installed behind the main mirror draws through a steady flow of air, preventing tube currents from building up and disturbing the seeing.

contrast, but it is not always straightforward and will vary from telescope to telescope. An online search will help you choose the best location and size for the baffle.

Alan MacRobert has used a novel method of blackening the inside of his 320 mm (12½-inch) reflector tube (*Sky & Telescope*, December 1992, page 696). He painted the inside with glue, on which he scattered handfuls of sawdust. When it was dry, he sprayed the whole thing with matt black paint, taking four cans to complete the job. He suggests that a more efficient method would be to dab the paint on with a sponge on a stick. As a result, the sawdusted area was truly black, though the lower parts of the tube, painted only, were still shiny. If you have an open or skeleton tube to your telescope, enclosing the tube in black plastic sheeting will be a good start, though this is reflective and not the ideal material.

Rich Combs, who observes from San Francisco's Bay Area, advises that to approach refractor performance with a reflector it is also necessary to attach a flat window at the top of the tube. Ordinary flat glass rarely has flat and parallel sides, so this glass needs to be optically worked to eliminate distortions, though a cheaper alternative is to use very thin monomer film, which is essentially Baader AstroSolar™ filter but without the coating. Baader claim that their TurboFilm™ introduces no more distortions than high-quality polished optical windows. A window on a reflector is advantageous for several reasons: first, it will cut down the currents that swirl around inside the tube when the telescope is not at ambient temperature; second, it will prevent dust from getting on to the main mirror; third, if it is a glass pane located just above the eyepiece (and well inside a dew shield or tube extension) it will allow you to do away with the vanes of the assembly that carries the secondary mirror. You can fix the secondary assembly directly to the inside of the window, and then adjust the angle of the secondary by slight movements of the window. Rich sums up:

> "In general, I would say for visual use, about 80% of the improvement you will notice in the scope's performance will be gotten from the first 20% of the effort you put into limiting stray light. This involves things like extending the input end of the tube, lining the interior of the tube with a photon-sticky substance, and closing the mirror end of the tube. The extra effort of fully baffling the scope will get you most of the other 20%."

Designs for better contrast

One source of low performance in reflectors is the size of the central obstruction caused by the secondary mirror. Not only does it obstruct some light, it also degrades the image somewhat through diffraction effects and reduces the fine contrast of the image though not the overall contrast. As shown in Figure 5.8, the image of a star consists of a tiny false disk, surrounded by spikes (and, in larger instruments, rings) produced by the diffraction of light as it passes through the circular aperture of the telescope, whether it be a refractor or a reflector. The effect of the central obstruction of the reflector is to throw slightly more light into the

▶ Fig. 5.8 The bright star Capella photographed through a 150 mm (6-inch) f/4 Newtonian. The spikes on the image are a result of diffraction caused by the vanes of the spider holding the secondary mirror.

spikes and slightly less into the disk. If you imagine a planet as a multitude of point sources of light, then a perfect image of it would reproduce each of those points perfectly. But if the image of each point is a tiny disk surrounded by diffraction spikes, this will cause a very slight smearing of the detail. However, Horace Dall, whose knowledge of telescopes was unrivaled, maintained that as long as the diameter of the central obstruction was less than 20% of that of the main mirror, the resultant loss of microcontrast was negligible. Nevertheless, the conventional SCT has a central obstruction of over 30% (see Figure 5.9), and the loss of fine contrast compared with an *f*/8 Newtonian is noticeable. The design with the smallest central obstruction, however, is the Maksutov-Newtonian, which uses a Maksutov corrector plate and a Newtonian secondary so you view at the side instead of at the bottom of the tube (Figure 5.10).

It is possible to reduce the size of the secondary in Newtonians at the expense of field of view, in order to cut the contrast loss to a minimum. If you are a planetary observer and have a good driven mounting, a wide field of view is unnecessary. Deep-sky observers could do this too, for many deep-sky objects are small, but if the field of view is reduced to a minimum it will be very hard to locate the objects in the first place.

The vanes, or "spider," which support the secondary mirror also cause diffraction effects that rob the image of some contrast. These are the source of the cross-shape in the images of bright stars, particularly in photographs. One solution would be to make the vanes very thin, but this can make the secondary subject to vibration. Another approach is to use elaborately curved vanes so as to spread the effect over an area instead of concentrating it into a cross.

With a large enough telescope, however, you can avoid the problem of the central obstruction altogether. A 300 or 350 mm (12- or 14-inch) reflector will have an

◀ *Fig. 5.9 A Schmidt-Cassegrain telescope has a large central obstruction – in this Celestron C14, it spans more than 30% of the aperture. This reduces the contrast of the instrument's images compared with those from a standard Newtonian of similar aperture.*

▶ *Fig. 5.10 An Intes-Micro MN78, a 180 mm (7-inch) Maksutov-Newtonian with a central obstruction less than 15% and performance similar to a similar-sized apochromatic refractor. Notice also the numerous baffles inside the dew cap and tube.*

appreciable gap between the vanes of its spider. Make a mask for the top end so that you look through only one quadrant, and you will have an unobstructed aperture of just under half the original mirror size. Since the seeing is often better with a smaller aperture, you may paradoxically be rewarded with sharper, though dimmer, images of the planets than with the full aperture.

Devotees of unobstructed-aperture reflectors use what are known as *tri-Schiefspieglers.* These have three mirrors, two spherical and one flat, which result in a totally unobstructed yet color-free image, with long focal length. Such instruments tend to be made to special order, but they are renowned for their contrasty images.

In the quest for contrast the eyepiece should not be overlooked. The trend today is toward wide-field eyepieces made of many glass elements which are usually coated to improve their light transmission and contrast. There are older and simpler designs which give particularly high contrast, such as the monocentric, but these have the drawback of giving an unfashionably narrow field of view. Nevertheless, the wide-field eyepieces are noted for their exceptional performance on the planets, particularly at short focal lengths. Lanthanum eyepieces contain special glass to produce very good *eye relief*, even at short focal lengths. Eye relief is the distance from the eyepiece your eye must be for you to see the full image. With short focal lengths this can be notoriously small, while an eyepiece with large eye relief can be used even by wearers of spectacles.

Whatever eyepieces you use, you will lose performance if they get dirty. I am as guilty as the next observer of leaving eyepieces lying around during an observing session to collect dew and then dirt. They should be regularly cleaned using proprietary lens cleaner and lens tissues.

To help you evaluate the effects of these improvements to your system, I have provided a set of charts of the stars near the north celestial pole which you can find at www.stargazing.org.uk/nps.

6 · CHOOSE YOUR AMMUNITION

This chapter covers the various technological aids, from new cameras to filters, that you can use with your telescope to make life easier for urban observing. During the 21st century, much of amateur deep-sky observing has become increasingly technical with the advent of CCDs. But this is nothing new. Since the mid-19th century, observers have been using a technical fix to reveal fainter objects than they could see by eye alone – photography. Even though it is now totally electronic in operation, and tends to be called imaging, photography continues to play an vital part in the amateur astronomer's armory, from the city just as much as in the country.

In the first edition of this book, written in the early 1990s, digital imaging was in its infancy. Not just the detectors but also computer storage were very limited, and the information carried by a small rectangle of film was hard to beat using electronic means. Today, however, the situation is totally different. Instead of random collections of photographic grains, our images are made up of rigid arrays of *pixels*, or picture elements. The CCD or CMOS chips that collect images can now often record just as much detail as most 35 mm film, with much more convenience, and memory cards smaller than a single 35 mm frame can store thousands of images. Many people have phone cameras that far outperform the compact cameras of the 20th century, though there are still a few cases where film does have its advantages. Furthermore, the digital revolution has meant that from city skies it is now possible to observe objects that have been totally invisible to visual observers in the same location for decades.

In astronomy, the traditional advantage of photography over observing at the eyepiece is that while the eye records only the light falling on it at any one instant, the imaging chip, and previously film, can carry on recording light for as long as you continue the exposure, building up an image. This is what makes it possible to photograph objects which are far too faint to be seen by eye. There is also the advantage that the image is recorded permanently, and can then be measured.

But photography need not be used solely to make a scientific record. Even with the simplest of cameras, from your city or suburban location you can take attractive and interesting astronomical photographs. There are some excellent books on *astrophotography*, as it is called, so here I'll just outline the sort of things you can do. If you want details of how to attach your camera to a telescope or what sort of mount to use, look in one of those other books. In practice, trial and error is usually the best guide, but I suggest exposure details that you can start with.

Using simple cameras

Modern digital compact cameras and even phone cameras produce brilliant images of everyday scenes, and to a limited extent they can also be used for simple astrophotography. One crucial factor is the maximum exposure time that the camera can give, which in the case of compacts is often limited to a matter of seconds whereas astronomical exposures are often longer than that. Another is the degree to which the camera's functions, such as focus, sensitivity and firing of the flash, can be controlled manually. The same applies to more advanced cameras, such as DSLRs (digital single-lens reflex cameras, with interchangeable lenses), which are also capable of more demanding photography.

So check in your manual how to switch off the flash, how to set the camera to focus on infinity, how to set its sensitivity, and how to over- or underexpose your shots. The latter are controlled in units known as *stops*, equivalent to the click stops on lenses, and the control is usually indicated by a plus-and-minus symbol. Each stop is a factor of two in exposure.

The ISO setting of sensitivity is equivalent to the values for film, and low values are often called "slow" and high settings "fast." The same applies to lenses and indeed telescopes. An f/2.8 lens is fast, while f/8 is slow. The lowest ISO setting on cameras is usually ISO 80 or 100, and a doubling of the figure means that you can halve the exposure time or close the lens down by a stop. So if your correct exposure is 8 seconds at ISO 100, you could also give 4 seconds at ISO 200, or ½ second at ISO 1600. Most compact camera lenses are around f/3.5 for the budget models or f/2 for the more advanced ones.

With manual camera control, and with the camera on a tripod (virtually every camera has a tripod socket), you can start to take sky photos, from pictorial twilight shots to constellation photos, which could well show stars fainter than you can see by eye, aurorae if any are around, and maybe even the occasional comet.

Twilight and constellation shots

The easiest pictures to take, and often the most aesthetically appealing, are simple shots of astronomical objects in the twilight. A crescent Moon, particularly with Venus or another planet nearby, can often be photographed with an exposure time of a second or so, even using the auto setting. You must keep the camera rock-steady. Holding it firmly on a wall is good enough for a 1-second exposure, but for longer than this a tripod is more or less essential. Far from avoiding foreground objects such as buildings or trees, make full use of them: an attractive

▲ Fig. 6.1 Venus is next to St Chad's Church in Shrewsbury, Shropshire, in this twilight shot taken with a compact digital camera. The exposure time was ¼ second at ISO 1600, so the camera was rested on a wall.

foreground will enhance the picture's interest. An example is shown in Figure 6.1.

One feature of digital cameras on their Program or Auto settings is that they will adjust their sensitivity automatically to cope with dim light, whereas with film you were limited to whatever you had in the camera at the time. So you can take one shot in brilliant sunlight by day, and the next shot in late twilight, and both can be perfectly exposed without requiring a very long exposure time for the night shot. Quite often you can take excellent twilight shots later in the evening than you might think, because the camera, if held steady, will boost its amplification to take a good shot.

You can photograph the setting Sun using your camera's telephoto setting without the need for a solar filter when it's low enough for you to look at it directly in safety. Photographs of the Sun setting between buildings or on a distant horizon with objects silhouetted against it are often well worth taking and turn a city location to your advantage. Getting the exposure right, however, calls for some care. The camera's exposure readings are generally programed to average the bright and dark parts of a scene, often with a weighting which assumes that the center of the frame is more important than the fringes. So you could get a bright Sun with black foreground, or an overexposed Sun with an unappealing gray foreground, depending on the circumstances. You could manually adjust the exposure setting to allow for this, but many cameras have the useful feature that if you partly press the shutter button, you fix the exposure setting and indeed the focus. Keeping the button partly pressed you then reframe the picture and take the shot, with the exposure and focus remaining the same.

However, if you want to show detail on the Sun's image, such as a naked-eye sunspot, then even the average exposure will be too great unless its image almost fills the field of view. You must underexpose the picture by at least two stops compared with what the meter reading suggests in order to show any sunspots. The same applies to attempts to photograph the green flash.

Later in the twilight you can photograph the first stars beginning to appear. You can allow the camera's exposure control to adjust the setting, but again if you get washed-out results you might need to take manual control. The best results come from a low ISO such as 200 and a longer exposure time, up to 10 seconds on a wide-angle setting, so as to give a smooth noise-free sky background (Figure 6.2). Even in late twilight the sky may may still come out an attractive blue, with stars visible, but in the presence of severe light pollution there is a point where the natural color becomes muddied.

You can even include streetlights in your view, as long as they are not too bright. If they are, you will probably get ghost images as a result of reflections within the camera. The same problem can arise if you include the Moon when it is anything larger than a thin crescent. Reflections inside the lens produce a reversed and fainter image diagonally opposite the Moon in the frame. This is more of a problem on high ISO settings. The only solution is to make sure that the Moon is exactly centered in the frame, which may not be very aesthetically pleasing.

▶ Fig. 6.2 This view of the twilight sky in 1996 from suburban Middlesex shows Mercury close to the horizon, the Pleiades star cluster at left and Comet Hyakutake at right.

Even at a low ISO, you may find that stars fainter than you could see with the naked eye show up. You may be tempted to give a longer exposure in order to bring out more of them. Try it and see. What will happen is that the stars begin to trail, more so if you are using a telephoto setting than wide angle. Look at your images at full size on the view screen to check for this when taking the shots.

So increasing the exposure time won't record fainter stars, though you might see them more clearly if they are short trails rather than just dots. By giving longer exposures you can create star-trail shots, and by stacking the images afterward in image-processing software you can combine the star trails without actually increasing the contribution from light pollution. (The trick, in Photoshop, is to copy each image into a separate layer and then change the mode for each layer to "Lighten" before flattening the image. Alternatively, set the mode to "Screen" to add the brightnesses instead.)

In order to record fainter stars with a fixed camera you must use a higher ISO setting such as ISO 1600 or 3200. This will reveal stars below naked-eye visibility with exposures of 8 seconds at f/4 in a dark sky, though at these high sensitivity ratings the results will be

◄ Fig. 6.3 High or low ISO?
The more you increase the ISO, the worse the electronic noise becomes. The top picture was taken with a Canon 40D at ISO 100 using an exposure time of 30 seconds, while the lower one was at ISO 1600 and an exposure time of 4 seconds. The two exposures are equivalent, but enlargements of the images (insets) show that while the stars are trailed on the 30-second exposure the sky background is smooth, whereas on the 4-second exposure the stars are points but the noise is much worse. The overall color is also muted. More recent cameras perform better at high ISO, but low ISO always gives smoother results.

▲ *Fig. 6.4 This shot of Auriga and Perseus over the estuary of the River Blackwater in Essex, UK, was taken with a full Moon in the sky. It was a 6-second exposure at ISO 1600. Moonlight turns the sky blue, just as sunlight does, though the intensity is too low for the eye to see the color well. But moonlight is slightly redder than sunlight, as the daytime view in the inset shows.*

more noisy. But from the city, again, the results will be less impressive because of the light pollution. There will be a noticeably colored or dense background, particularly near the horizon, as in Figure 6.3. Nevertheless, if you wait for those really clear nights and concentrate on the area around the zenith, you will avoid the worst effects of light pollution.

Oddly enough, it can be more rewarding to take pictures of a brightly moonlit sky. Moonlight reproduces a delicate blue on high settings, and can overpower average light pollution. Particularly faint stars do not show up, but the results can be pretty. Even thin streamers of cloud can lend a romantic look to the picture. Color is also revealed in the landscape, and in country locations with no lights visible a moonlit landscape can look just like a sunlit one, except that you can see star trails in the sky (Figure 6.4).

Pictures taken in moonlight or even light pollution are quite useful for depicting the constellations. Because only the brighter stars are recorded, the main patterns are easier to see. Some astrophotographers deliberately use a slightly misty filter to spread out the star images slightly, which both brings out the stars' colors and makes the constellation patterns more recognizable.

Driving the camera

Having exhausted the repertoire of pictures you can take with a fixed camera, your next step is to drive the camera so that it tracks the stars. This overcomes the problem of star trailing, and allows you to make longer exposures. Even compact cameras, which are often limited to a maximum exposure of 30 seconds, could be used to photograph fainter stars in this way. With a camera that has a B setting (which allows you to give unlimited exposure times) you'll need a cable release of some sort, which holds the shutter open, or alternatively to link the camera to a computer using the appropriate software to give time exposures.

From now on, compact-camera owners have a harder time, because most astrophotography involves not only long exposures but also the

◀ *Fig. 6.5 The AstroTrac is a portable camera platform which uses a screw thread at the end of long arms to simulate a very large worm wheel. It is accurate enough to allow driven exposures with telephoto lenses or even lightweight telescopes, and the pole-finder allows you to get going within minutes.*

▼ *Fig. 6.6 The Andromeda Galaxy, photographed with a 200 mm lens on a Canon 7D DSLR mounted on an AstroTrac TT320 from the dark skies of Lapland. This is a stack of five subframes with a total exposure time of 4m 49s.*

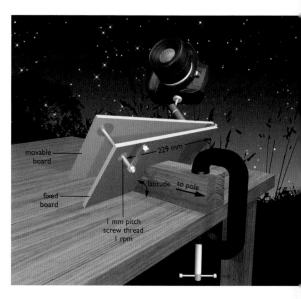

▶ *Fig. 6.7 The Scotch mount is a simple means of driving a camera to follow the stars. This version is designed for northern hemisphere use – the hinge would be at the other end of the boards for use in the southern hemisphere. A simple sighting device parallel to the hinge will help you to align the mount on the Pole Star with an accuracy good enough for exposures of up to 8 minutes with a standard lens.*

ability to either change the lens or to replace the lens by a telescope. This usually means a DSLR camera, though what are called CSCs (compact system cameras), which also have interchangeable lenses, can also be used, with limitations.

The main requirement for exposures driven to follow the stars is a driven equatorial mount of some kind – that is, a mount with one axis fixed parallel to the Earth's axis. Altazimuth mounts have the drawback that they can't deal with the rotation of the image as the object moves through the sky. You can buy or make a battery-driven *camera platform*, which is a mount that will drive the camera at the correct rate. The very portable but accurate AstroTrac (Figures 6.5 and 6.6), the iOptron SkyTracker and the Vixen Polarie are examples of such devices. They are particularly useful for the urban astronomer who doesn't want to bother with, or have room for, a fully fledged telescope and mount yet wants to be able to take guided photos when on vacation in dark skies.

It is also possible to make a simple screw-driven mounting known as a *Scotch mount* or *barn-door mount*. This consists of a pair of hinged boards or plates, mounted so that the line of the hinge points toward the celestial pole. A screw or bolt threaded through one board, which is fixed, pushes against the other board and moves it slowly as the screw is turned (Figure 6.7).

The trick is to make the device in such a way that turning the screw at the rate of one revolution per minute – in synchronism with the second hand of a watch – drives the movable board at the rate of one revolution per 24 hours. In practice, of course, a full revolution is physically impossible, and after 10 minutes or so of driving the rate becomes inaccurate, but that is about the limit of endurance for hand driving anyway.

The Scotch mount offers a very cheap means of taking driven exposures, but for city astronomers its usefulness is limited to fairly short exposures with standard focal lengths, between the wide-angle and telephoto settings. The errors in the drive will become obvious if you try to use longer focal lengths. With a standard lens you should aim to turn the screw so as to keep it within 10 seconds of the position of the second hand. With a telephoto lens, the accuracy should be within 3 or 4 seconds of the second hand, but with a wide-angle lens you can afford to be less precise. Using ISO 400 and a lens of any focal length at $f/4$, exposures longer than 30 seconds will start to be seriously affected by light pollution. With a lower ISO or slower lenses, exposure times can be longer. An exposure of 4 minutes at ISO 100 with a telephoto lens setting would require extreme care, but could give good results even in suburban conditions.

Even if your lack of a drive limits you to 30-second exposures, you can still get the effect of longer exposure times by stacking the images in photo-editing software and registering the star images. This brings out the fainter star images which otherwise would be lost in the noise of a single image. The free Deep-Sky Stacker will do the job for you, and will allow for any slight rotation of the image as a result of misalignment on the pole.

If you have a telescope on a motor-driven equatorial mount, you can often mount the camera piggyback on top of the telescope (Figure 6.8). The cradle that holds the telescope tube frequently carries a camera thread just for this purpose. Even the simplest electric motor drives are good enough to allow exposure times of several minutes with a moderate telephoto lens, assuming that your polar alignment is good. There's more on motor-drive accuracy on page 139. The AstroTrac motor-driven camera mount offers very good driving as well, and some people use it in place of an equatorial mount for their small telescopes when traveling.

However you drive your camera, you will soon find that from the city and suburbs the limits to exposure are soon reached if you are using just a standard lens. Even a low ISO will start to pick up the sky background after a few minutes. Slow settings have the advantage that, although you need to expose for longer to record the same image, the contrast of the image is generally higher, and the noise lower, than

▶ *Fig. 6.8 The tube rings of many telescope mounts have a thread that allows you to attach a camera, so you can use the telescope's drive system to take long exposures.*

for high settings. Images of stars and nebulae are therefore easier to pick out against the sky background. You will have to experiment for yourself to discover the best exposure time and ISO on a given night from your location. It is surprising how many stars you can record in this way, even from a light-polluted location. Objects such as the Orion Nebula and the Double Cluster in Perseus show up well, and even the Milky Way can begin to appear at the right time of year from the outer suburbs or smaller towns, particularly with a bit of image processing. One trick, which doesn't actually reduce the light-pollution level but makes it a more acceptable color, is to set the camera to its tungsten-lighting (incandescent) white balance rather than daylight or Auto. This makes the sky go a more attractive neutral or blue color (Figure 6.9).

So far I have stuck to talking about standard lenses, which means around 35 mm focal length for DSLRs. But, as with visual observing, increasing the magnification for a given aperture darkens the sky background more than it does star images. You could simply stop down your lens from, say, $f/3.5$ to $f/5.6$, but it is more appropriate to use a lens of longer focal length, one with (usually) a larger f-number than the standard lens.

Bear in mind that to record a faint star you need more aperture, but to record a faint extended object, such as a nebula (or the sky background), you need a lower f-number. Most people do not appreciate that for a given exposure time and ISO you can record just as much of the extent of, say, the Orion Nebula using a standard lens at $f/3.5$ as you could using a giant telescope also working at $f/3.5$. The difference is that the big telescope, with a mirror several meters across and greater focal length, will have a much larger image scale, with the nebula covering maybe 150 mm rather than 0.5 mm or so, and will

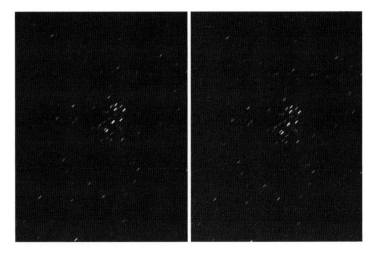

▲ *Fig. 6.9 Two identical 30-second star trails of the Pleiades star cluster. At left, using daylight white balance, and at right,* *using the tungsten white balance. The results are very similar to correcting out the light pollution by image processing.*

therefore show much finer detail. And, of course, from the suburbs the fainter extensions of the nebula will be masked by the light pollution.

With lenses of longer focal length, the driving requirements for your mounting and its alignment are more demanding, but with, say, a 200 mm *f*/4 lens and ISO 400 on a DSLR you can achieve excellent results from urban skies with exposures of two or three minutes or even longer. This is a typical maximum zoom lens focal length. The aim is to find an exposure time that will produce images with a just-acceptable amount of sky background. This will vary widely from night to night, depending on the clearness of the sky. I find that the camera's viewscreen can sometimes give a brighter view of a dark subject than the downloaded image, so it's worth checking the histogram, usually accessible by pressing a button alongside the view screen.

This shows a graph of the brightnesses in the image, with black at the left and white at the right. Ideally, your exposures should have a peak somewhere in the left third of the histogram, which will give you plenty of information to process. If the peak is very close to the left edge, you probably don't have enough exposure, and you should try to get the foot of the curve of the peak some way from the left-hand edge. But if it is flat for some distance along the bottom at the left before it starts to rise to a peak, you can reduce the exposure time as you are simply adding sky brightness rather than gaining useful information. Similarly, a histogram that has a flat right-hand edge

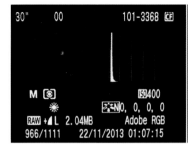

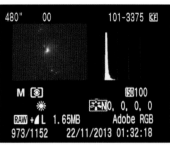

▲ Fig. 6.10 Use the histogram on your camera viewscreen to check your exposure. The left-hand shot is underexposed as the histogram is too far to the left and has a sharp left-hand cutoff. The right-hand one is much better as it carries all the information you can use. Don't be afraid to have some bright sky.

is losing information because you are saturating the brightest levels (Figure 6.10).

With this sort of setup you can photograph a range of open clusters, the brighter nebulae and some galaxies. You should be able to take attractive pictures of the Orion Nebula, for example, which may encourage you to aim for other objects. However, the Orion Nebula is so much brighter than any other nebula that few others will show up anything like as well. Its only real photographic rival is the Eta Carinae Nebula (NGC 3372) in the southern hemisphere. The Andromeda Galaxy should photograph as an elliptical glow, but the fainter spiral arms may not be revealed. Fainter galaxies such as the Whirlpool Galaxy (M51) may not show up on individual frames but you should be able to bring them out with some image processing. The image scale of a 200 mm lens is too small for it to reveal much of globular clusters, which will appear as small hazy blurs. Even

▶ Fig. 6.11 In 2007 Comet Holmes brightened up spectacularly from 12th magnitude to naked-eye visibility. This shot was taken with a 200 mm lens and an exposure time of 210 seconds at ISO 400 on a Canon 40D camera. Image processing has removed the light pollution.

Omega Centauri, the largest and brightest globular, is not very large when photographed with a 200 mm lens.

The same applies to most Solar System bodies – the Moon is only 2 mm across on the chip when photographed with a 200 mm lens – so I'll deal with the Sun, Moon and planets separately since they really need to be photographed through a telescope. But comets are another matter, and sometimes the brighter ones with a tail are too large to fit within the field of view of a telescope, so a telephoto lens is ideal (Figure 6.11).

Photography through the telescope

It's possible to use a compact camera to photograph through a telescope by simply pointing the lens into the eyepiece, ideally using a commercially available clamp for what is known as *digiscoping*. You can get good pictures of the Moon (Figure 6.12), and maybe passable pictures of the bright planets in this way, but giving even 30-second exposures of deep-sky objects starts to become problematic. For the best results, you need to be able to remove the camera lens and replace it by the telescope, using either a DSLR, a CSC, a webcam or in some cases a specialist video camera. Each has its own uses.

Using a telescope for photography is nothing more than using a very long telephoto lens. If you put your DSLR or similar camera, minus lens, in place of the eyepiece of one of the popular 200 mm (8-inch) Schmidt-Cassegrain telescopes (SCTs), for example, you are effectively using a 2,000 mm f/10 telephoto lens. This setup will give an image of the Sun or Moon about 18 mm or ¾ inch across (Figure 6.13), which is rather larger than the common APS sensor size used on most DSLRs (around 22×15 mm, though it varies from camera to camera). This image scale is large enough for photographing a wide range of

▶ *Fig. 6.12 I took this daylight shot of the gibbous Moon through 12×45 binoculars using a cheap digital camera held up to the eyepiece. The closer the Moon is to full, the brighter its surface, so the easier this becomes.*

▶ *Fig. 6.13 The third-quarter Moon photographed using a Canon 40D at the prime focus of a Meade LX90 Schmidt-Cassegrain telescope, which has a focal length of 2,000 mm.*
As soon as the phase exceeds a fat gibbous, the Moon's disc is too large to fit fully within the field of view.

objects. In general, telescopes have focal lengths between about 500 mm up to 2,000 mm, and each type has its own merits. The shorter ones are good for wide-field nebulae, star clusters and bright comets, while the longer ones are more suited to small galaxies, planets and planetary nebulae. All these objects are within reach of city astronomers, though it is probably better to go for the longer focal lengths and to concentrate on planetary, lunar or solar photography.

Each type of photography has its own technique, of course, with some crossover between them. I'll start with what is really the digital version of traditional photography, which involves using a DSLR camera, then move on to the more esoteric versions using specialist cameras.

Connecting cameras to telescopes

The standard means of connecting a camera to a telescope is to get what's called a T-ring or T-adapter with the specific bayonet fitting for your DSLR camera. The T-adapter is a standard photographic lens adapter with a 42 mm barrel having a screw pitch of 0.75 mm, and many telescopes already have this thread either on the focus mount or on another adapter (Figure 6.14). This is slightly different from the old 42 mm Pentax screw thread with a 1 mm pitch used on a lot of old SLR lenses. So if you have a thread that looks as if it should fit but doesn't quite do so, you are probably trying to screw these two incompatible threads together. Most specialist telescope suppliers have T-adapters to fit all common cameras to their telescopes, or if all else fails you can get a standard 1¼-inch eyepiece barrel with T-thread, which has the advantage that you can then fit your camera on to virtually any modern telescope.

Some beginners' telescopes do not have enough inward travel on the focuser to bring a DSLR to focus. The focal plane on a DSLR is some 45 mm behind the lens flange (it varies from make to make). This distance is sometimes referred to as the *back focus*. The only way you could focus a DSLR on such a telescope might be to use a Barlow lens or eyepiece projection. This is also a common technique for increasing the focal length, so you might need to do this anyway when imaging planets.

The technique is to use the Barlow lens between the telescope and the camera. The increased effective focal length should bring the focal plane out just far enough for you to focus the camera. You might think that the new focal plane would be a long way from the camera, if you have doubled the focal length, but this is just effective focal length. The light emerging from the telescope appears to have the longer focal length, but the focal plane is now just a short distance farther out. In the case of eyepiece projection, you'll need an extension tube with provision for an eyepiece to be placed midway along it, but in this case no lens on the camera. With the right eyepiece you may be able to reach focus with a much enlarged image, though in my experience many eyepieces either won't fit or don't give the required focus position. A CSC has a much smaller distance between the lens flange and the focal plane, so such a camera might come to focus in cases where a DSLR won't, and is also much lighter which will make it easier to balance the telescope on the mount. Webcams and CCDs also have a focal plane closer to their mounting flange, so may also reach focus (Figure 6.15).

You may encounter the opposite problem, where the focus point is too far out for the focusing range. This can happens with refrac-

◀ *Fig. 6.14 Many telescopes have T-threads on their eyepiece mounts which allow you to attach almost any camera that takes interchangeable lenses using a widely available adapter.*

▶ *Fig. 6.15 The focus point on some instruments – here, shown as a cone of light on a Sky-Watcher 130 mm reflector – may be too close to the focusing mount to allow room for the back focus of a DSLR.*

tors, not necessarily cheap ones, which are designed to be used only with a star diagonal, though often the extra focal plane distance of the DSLR is sufficient by itself. Otherwise, you may prefer to use an extension tube instead, to avoid the image inversion caused by the diagonal.

If you plan to buy a telescope for astrophotography, it is worth checking with the supplier in advance that it has a suitable focusing range for whatever camera you have in mind.

Photography using a DSLR or CSC

This is the starting point for many people, as DSLRs are the standard for high-quality everyday photography as well. With a DSLR camera, you can view the image it is taking through either its optical viewfinder, which views directly through the lens, or in most recent cameras using the viewscreen on the back of the camera (called *live view*). The optical viewfinder uses a pentaprism, which adds to the bulk, but a new breed of camera, known generically as the compact system camera (CSC), does away with this and gives the advantages of interchangeable lenses with much reduced size. Generally, they have reduced photographic features as well, though as long as they have the facility for manual control and have a bulb setting for the shutter, they can be used in place of a DSLR for basic imaging. Most of the following applies to CSCs as well as to DSLRs.

There are actually some DSLRs specifically designed for astronomical work. One of the first things that astrophotographers noticed when they started to use digital cameras was that their response to the deep red color of nebulae was much worse. Although camera chips have strong sensitivity to infrared (IR) light, it is reduced in commercial cameras by an infrared blocking filter directly in front of the chip in order to improve the color rendering, notably of skin tones. Too much infrared sensitivity and people's faces go very blotchy. For some reason the camera manufacturers have chosen to filter out more IR than that which was perfectly acceptable with film.

In the past, it was noticeable that a film photo of the Orion region, even star trails, showed the Orion Nebula as a bright red spot. Even in light-polluted areas it was not difficult to photograph quite faint nebulae such as the North America Nebula or even Barnard's Loop, a huge feature near Orion. But with digital, these objects are harder to image, even from country areas.

It is possible to modify a DSLR to remove this filter, but it isn't as easy as simply removing the lens and unscrewing it. The chip is buried way down in the innards, and to remove it you need to know the exact procedure for dismantling the camera, detaching all the delicate foil connectors and, what's more, getting it all back together again afterward. The instructions are available online but you need either nerves of steel or complete recklessness to try it. I know – I've done it (successfully!). And simply removing the filter isn't enough – you need to replace it by a non-blocking filter of equivalent thickness if you want to preserve the focusing accuracy on the viewscreen. There are people who will do this for you if you don't want to risk your own abilities (Figure 6.16).

The alternative is to buy a camera which already has this modification. Canon manufacture a version of their popular EOS 60D model, the 60Da, which not only has enhanced sensitivity to hydrogen light but also has greater chip sensitivity. This follows on from their earlier EOS 20Da. You can use the camera for everyday photography by choosing a different picture-taking mode, so it is not useful purely for astronomy. Bear in mind that it has somewhat limited usefulness – it will not give vastly better results of galaxies, for example, except to reveal nebulae in spiral arms.

Once the camera is attached to the telescope and you know you can focus it on astronomical bodies, you can begin taking some test exposures. The best object to start with is the Moon, which has the great advantage that it is around every month, is easy to find, and is bright enough that exposure times are usually only a fraction of a second. Its size and brightness make it by far the easiest Solar System body to photograph, but it's also excellent practice for moving on to planetary photography. After all, planets are only the same apparent size as individual lunar craters. More details are given below.

With the planets, the problem of image scale starts to become obvious. The largest planet in terms of its image size is usually Jupiter, which can get up to 50 arc seconds in diameter when at its closest, compared with over 2,000 arc seconds for the Moon. With most telescopes the image is very small in the frame, and while you should be

▶ *Fig. 6.16 M42 photographed with the same Canon 10D camera, with the infrared blocking filter present (top, 45-second exposure) and removed (bottom, 30-second exposure). Removing the filter records the hydrogen-alpha emission more strongly.*

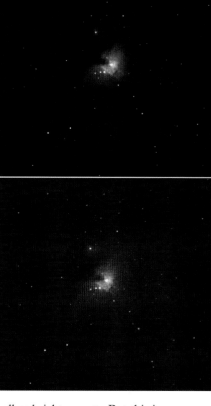

able to photograph its main dark belts with an exposure time of a fraction of a second, there won't be much detail there. The other planets are even smaller.

This is where a Barlow lens or eyepiece projection is essential. Using such a system you need to increase your effective focal length to many meters, which will then give you a tolerable image size. This will also allow you to photograph small areas on the Moon. However, serious planetary imaging these days is the preserve of the webcam, or rather the webcam-type camera, covered later in this chapter.

A DSLR is more appropriate for photographing the larger deep-sky objects, such as nebulae, star clusters and nearby galaxies, as well as bright comets. But this is where you could come up against another problem – errors in your telescope drive. Additionally, if you have an altazimuth mount, there is also the difficulty that the field of view will rotate, even if the central point is tracked perfectly. Field rotation varies across the sky, but in general it begins to show up after an exposure time of 30 seconds or a minute. However, if you can keep your exposure times short, and add the subexposures during image processing, you can to some extent overcome the problem.

Owners of equatorial mounts are not much better off to begin with. A few very expensive telescope mounts are capable of driving perfectly for many minutes on end without the need for any correction at all, but most aren't. There are two major reasons – bad polar alignment, and periodic error in the drive system. The first is

◀ Fig. 6.17 The worm-and-wheel mechanism of a Meade LX90 declination drive. Any slight error in the alignment or mesh of the RA gears will result in periodic error in the drive rate with every turn of the worm. The LX90 has periodic-error correction to help correct for this.

down to the care with which you set up your mount, and even an inbuilt pole finder is not always accurate enough to avoid it. If you can refine your alignment over time, and leave the mount in place, you can overcome it.

Periodic error is common to many mounts, though often there is correction software built in to the more advanced handsets. Telescopes are usually driven by a worm-and-wheel arrangement (Figure 6.17), and any slight misalignment between the two will result in a star image, say, moving slowly back and forth as the worm rotates, with a period of several minutes. This is unnoticeable visually, but becomes obvious when you do exposures of more than a minute or two.

The simpler drives have handsets with only limited controls – basically speed up, or slow down, for moving around a field of view. But more advanced mounts allow much more refined speed control, with fast slew and, usually, Go To facilities. These are the ones that offer periodic-error correction, or PEC. To make use of it, however, you need either a guiding eyepiece that contains crosswires, or a webcam. The procedure is to monitor the periodic error for one whole rotation of the worm while the software records your corrections. Then, as long as the position of the worm is known, the software can play back the corrections automatically. It sounds fine in principle, but in practice it is tedious and prone to error.

Instead, most people these days plump for autoguiding. This involves using a separate guiding telescope and CCD camera or webcam to monitor the drive rate of a drive system which includes an autoguider port. Typically you need to run the operation using a laptop computer near the telescope. This all sounds straightforward, but it is not usually plug-and-play. The cables supplied by mount manufacturers to link their mounts to your computer usually have the obsolescent serial (RS-232) connectors, which means getting an adapter cable, and the software that allows the mount to talk to the computer is produced by a third-party organization, ASCOM, and must be downloaded separately. Autoguiding software, such as the excellent free PHD guiding, is also needed, allowing you to watch the guide star and corrections on your laptop. The whole thing is not straightforward, and may require a lot of internet searching for help from people who have gone through it before. Standalone autoguiders are available, which don't need a laptop and display simplified star images on their viewscreen (Figure 6.18). Some people find them useful but they may have lower sensitivity than CCD autoguiders, requiring a larger guide scope.

An autoguider does not do away with the need for accurate polar alignment, but quite a basic system, using a fairly cheap webcam, is adequate for many purposes. And in urban skies, you may find that your sky is so bright that exposures longer than a few minutes or so are a waste of time anyway, so you could stick to short subexposures.

The more subexposures the better. You need to expose each one so that no part of the image you are interested in is so bright that it is washed out, so aim for a medium background sky level, following the same histogram guidelines as for general sky photography. Use as low an ISO as possible while still being able to guide satisfactorily for the longer exposures that you'll need to get each image well on to the histogram. You can't give 3,600 1-second exposures to get the same effect as a single 1-hour exposure – each subexposure has to get well above the sky background level.

Though you can use the DSLR as it is, which requires the minimum of effort, there are two things you can do to improve your chances of success. One is to run the camera from an external power supply to avoid the battery running down during an exposure. The other is to control it from a laptop. Canon DSLRs are sold with basic software and cables for doing this, but currently these are a fairly costly extra

▶ *Fig. 6.18 The Synta Synscan Autoguider is a standalone autoguider that, when attached to a separate guide telescope, detects stars in its field of view and provides feedback via the autoguider port to correct for errors in the drive rate.*

with Nikon cameras. The advantages of shooting "tethered," as it is known in the world of professional photography, are considerable. You can check the focus of your shots at full size on the laptop monitor, whereas the camera viewscreen may combine pixels and give a blurry result even for perfectly focused images viewed at maximum magnification. You are not limited by the capacity of the camera's memory card. And the software has the facilities for sequences, bulb and time exposures. Other proprietary software, such as APT (www.astroplace. net), offers further benefits such as allowing programed series of exposures. Finally, as long as your cables are long enough, you don't actually need to be outside but can control both the camera and the telescope from your computer indoors (Figure 6.19). CSCs at the time of writing do not have the facility to shoot tethered.

Astrophotography is an area where saving the raw image can be important. Normal images such as JPEGs contain 8-bit data – that is, there are 256 brightness levels for each pixel. But the raw images (which have different file describers depending on the camera make) are usually 16-bit images, with 65,536 brightness levels. When you are stretching an image during processing to get the faintest possible detail, 8-bit images can start to show the individual brightness levels whereas 16-bit images should be smoother.

I also recommend turning off the in-camera image sharpening, which is often present by default. This can result in unpleasant black circles around star images. You can always sharpen images afterward using

any image processing software, but it's virtually impossible to get rid of once present, though raw images should be free from it.

Once you have your subexposures, then you need to process the results into a usable image.

◀ Fig. 6.19 One advantage of shooting with your DSLR "tethered" to the computer is that you can extend the leads sufficiently to allow you to sit indoors while the exposures are being made. Most sky mapping software will also allow you to operate the telescope's Go To function from the computer, so as long as it is well aligned, and remains in focus, you can carry out your imaging in comfort.

Processing DSLR images

Be warned – processing astronomical images generally takes just as long as obtaining them in the first place, the only advantage being that you are sitting at the computer rather than outside. But without the ability to process the images, you are limited to their appearance straight from the camera, which is often very uninspiring. You need suitable software, of course. The industry standard is Adobe's Photoshop, which in its full form costs easily as much as a DSLR. But the much cheaper Photoshop Elements will do most of what you need. Some people prefer Corel's Paint Shop Pro, while GIMP is free software with many of the same capabilities. The crucial features are common to all the products. For simply stacking subexposures and making sure that the combined images are all in register with each other you can use free software such as Registax or Deep Sky Stacker, but these do not have the same degree of control for post-processing the brightness levels and color offered by the other programs. There are also all-encompassing astronomical software programs such as Maxim DL that will do the same job, though again at non-trivial cost. If you are venturing into the world of astrophotography for the first time, bear in mind the cost of software that you might need to get the best results.

To combine subexposures using a Photoshop-type program you need to add the frames as separate layers on top of a base layer, then choose a suitable layer mode for each one (usually Screen mode, which adds the brightnesses), or alternatively give each layer a set transparency before flattening the layers so as to give a single com-bined image which will have considerably reduced noise.

The main adjustments you need to make use the Levels controls, which show you the histogram of all the brightnesses within the image. Using sliders you can choose the brightest and darkest levels that you want to see in the image, and also where the mid-brightness point should be. So if the image is generally too bright because of light pollution, move the mid-level control so as to darken the image to a more normal level, then adjust the top and bottom points to match the darkest and brightest parts that you want to see. Bringing the top point down too far brings out faint detail close to the sky background level but also washes out the brightest parts of the image, so it needs to be used with care. In the full version of Photoshop, but not Elements, there is a Curves control which gives you more precise control of this, and which is particularly useful for maintaining the highlight detail in the nuclei of galaxies and the centers of bright nebulae. The Trapezium of stars in the middle of the Orion Nebula is a good example of a feature that is easily washed out, yet judicious use of Curves or the midpoint slider can retain it while still showing the outlying detail.

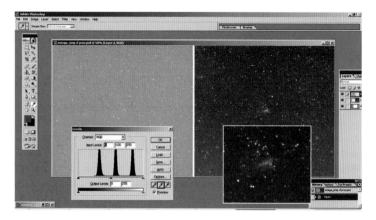

▲ Fig. 6.20 Basic image processing. A single shot of the Omega Nebula, M17, taken with a DSLR has a strong light pollution background (left pane). The Levels dialog box shows the histogram of the image, with separate red, green and blue components because of the background color. Click on the middle eyedropper (here shown outlined in red) and then on the sky background, and it immediately turns neutral gray, as shown in the right-hand pane. The inset on this pane shows an enlargement of the nebula area, which is quite noisy. Using many subframes will reduce this noise level to give a smooth background that will allow further enhancement.

The simplest way to get rid of the strong light-pollution color is to use the central eyedropper in the Levels box. With this highlighted, click on a part of the image that you think should be a neutral gray color – the sky background. Straight away the sky background turns neutral instead of orange. However, this is only a quick fix – in the presence of strong light pollution the bright stars may all go blue as a result, requiring more careful processing, so it is not always the ideal answer (Figure 6.20).

The reason for having so many subexposures is to reduce what is termed the *noise level*, by analogy with sound recordings. Each image has electronic noise, particularly at the higher ISO settings, which you can see as unevenness in the sky background as a random stippling of blob sizes and colors rather than being at a uniform level. The faintest stars or levels of actual brightness in the image are obscured by these much greater random fluctuations. Each image has different noise, so adding them together and bringing them all back to the same level means that the random noise will eventually be evened out. Then you can alter the levels to bring out the faintest parts of the image without exaggerating noise instead. But there will always be a limit, when the faintest areas are well below the sky brightness, and the brighter your skies, the sooner you will reach that limit.

The need to reduce the noise level is the reason for using a low ISO rather than the highest that your camera will give you. The high ISOs are fine for making sure you have the right target and for checking focus, but for actually taking good pictures you need to reduce the noise level, though the more recent cameras can give better results at high ISOs than older ones. One thing that people often don't realize is that the camera does get quite hot during long exposures, which increases noise. Those of you who live in warm climates have a penalty to pay for the comfort of your observing, as well as the mosquitoes – much more sensor noise! By contrast, we who shiver in northerly winters can at least get smoother images as a result.

There is far more to image processing than I can give here, but this is a start.

Photography using CCD cameras

A DSLR can only take you so far, particularly from an urban location. Out in a dark-sky area you can go on adding subexposures for as long as you please, and can get some excellent results. But the noise levels are usually a problem for those with light pollution to contend with, and for many people a specialist CCD camera is the answer.

These cameras usually cost much more than a DSLR, and look unlike any camera you are used to. There is no viewfinder, no shutter button to press, and of course no lens. Most modern cameras use the T-thread system to attach them to the telescope (Figure 6.21). A computer is essential, and usually also a 12-volt power supply. What's more, many of them take only mono images, with considerably fewer pixels than even the simplest DSLR or even phone camera. Yet they can take some amazing images, which are the rival of any taken at professional observatories not so many years ago. Their CCD (charge-

▶ Fig. 6.21 An Atik 314L+ CCD camera with a T-thread and nosepiece that will allow it to be fitted in place of a standard eyepiece. The camera has electronic cooling, with fins that help to dissipate the heat that is removed from the CCD itself.

coupled device) sensors are more sensitive than the CMOS sensors used in DSLRs, and furthermore they have provision for cooling, usually by means of an electronic cooling panel though some do require water cooling.

In many cases you have a choice of mono or color sensors, and one's natural inclination is to go for the color version. After all, we normally want nice color images anyway, and the alternative is very tedious by comparison. To get a color image from a mono chip you need to expose through separate red, green and blue filters and combine the results, making at least three times the work.

Color CCDs have the same sensors as mono chips, but the color filters are superimposed over the pixels, so some pixels are permanently filtered red, some green and some blue. The same array, known as a Bayer matrix, is used in most color digital cameras. The final color image has to be "debayered" so as to make an image in which all pixels are the color of the scene at that point (Figure 6.22). This means that the color information is only a quarter of the resolution of the image as a whole, but normally we don't notice this. It becomes more critical when we look at images of stars, however. On film, you could see the colors of stars well, but on DSLR images the smaller star images are often just a uniform white, partly because they cover too few pixels to contain enough color information. But deliberately defocusing or diffusing the DSLR image helps to reveal the star colors.

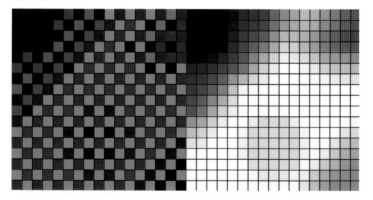

▲ Fig. 6.22 The Bayer grid is an array of red, green and blue pixels, with twice as many green as red or blue. At left is the pattern of pixels, with their brightnesses showing the output from each one in diagrammatic form. Debayering takes a group of four pixels and converts them to a color, then does the same for the adjoining group of four, producing a result which has intermediate colors though at lower resolution than a mono image, as shown at right.

The filters considerably reduce the sensitivity of the chip. To compensate for this you would have to give very much longer exposures, though of course if you are having to expose through three separate color filters anyway, the effort is comparable. But there are some significant advantages to using a mono camera. The greater sensitivity of the mono chip means that you can image faint objects in a reasonable time. You may think that you always want a color image, but armed with just a mono CCD you can observe nebulae and galaxies on your laptop that you could never hope to see visually from the city. And for color imaging, getting a good mono image is two-thirds of the battle.

As I mentioned earlier, the color information does not need to be as detailed. You can take the three filtered images at lower resolution by *binning* the color data. A group of four pixels are taken together, which improves the signal compared with the noise and gives a better image for the exposure time. It is effectively like doubling the sensitivity, so you can accumulate the color data quite quickly. As with DSLRs, the greater the number of subexposures you can do the smoother your final image will be. The unfiltered mono image is referred to as the luminance part of the image, and this provides all the detail and the brightness information. The color image by comparison can be of poorer photographic quality as long as the color information is there. The result is called an LRGB image, as compared with an RGB image where each filtered exposure is of equal quality.

I have tested a mono and color version of the same camera, the Atik 314L+, and the results were definitely in favor of the mono version, which was noticeably more sensitive (Figure 6.23). If you have reliably clear skies and accurate guiding, you should be able to get good results from the color camera, but there is much more versatility in having the ability to choose your color filters.

The R, G and B filters designed for astrophotography are carefully chosen so as to maximize the divisions between the colors. You also have the option of using narrowband filters, which are particularly useful for city astronomers. These isolate small parts of the spectrum where the gases in nebulae in particular emit strongly, such as the light of hydrogen and oxygen, but light pollution is weaker. In this way you can increase the signal-to-noise ratio in your images. You can choose the colors represented by these gases somewhat creatively, resulting in colorful images which are just as valid a representation of the object as the conventional versions.

With a cooled CCD camera and judicious choice of filters and exposure times, it's possible to take images of really quite faint objects. Bob Garner lives alongside the main A40 route into London, with strong

▲ *Fig. 6.23 Color or mono CCD camera? At left is a 30-second exposure of the lenticular galaxy NGC 3115 in Sextans, made with an Atik 314L+ color camera. On the right, the same object with the same exposure but with the mono version of the camera, which shows fainter detail. The instrument used was an 80 mm refractor.*

streetlighting literally within yards of his house, which fortunately is to the south of the road so he is looking away from them. With his 355 mm (14-inch) *f*/5 Newtonian reflector he regularly images galaxies of 15th or 16th magnitude (Figures 6.24, 6.25 and 6.26). "My main problem is not so much the light pollution as the aircraft," he says. "I get my best results in the early morning when Heathrow shuts down." His work is proof that a large instrument is still useful in the city, though even with this, total exposure times are often several hours.

Bob lives only about two miles from the site in Ealing where, in the late 19th century, A. A. Common took pioneering images of the Orion Nebula which were acclaimed at the time as being triumphs of the use of photography in astronomy, and which earned him the Gold Medal of the Royal Astronomical Society. Today, anyone with a small refractor and DSLR can far exceed the abilities of Common's 36-inch reflector and home-sensitized plates, but he would no doubt be pleased that astrophotography is still possible from the area (Figure 6.27). In Common's day, Ealing was surrounded by green fields and would probably have had darker skies than even most rural parts of England today.

Don't underestimate the effort required to take images, even in good skies. As well as getting the whole system to work satisfactorily, quite possibly setting it up from scratch each time you observe, there are many other aspects, such as the need to take dark frames (which allow for imperfections in the camera) and flat frames (which remove unevenness in the optical path and dust on the chip). These are also essential for the best results with DSLRs and other cameras.

But not only will CCDs bring you the chance to take images that you can be proud of from your unpromising location, they can also enhance and increase your personal observing opportunities. Those galaxies and nebulae that you read about – many of them can be yours from your light-polluted site with exposure times of maybe just seconds. You don't even need to save or process the shots – there can be as much fun in picking them out on your laptop as observing them visually. The number of objects accessible to the urban astronomer is multiplied many times by using a CCD.

Photography using webcams

Just as CCDs have transformed deep-sky observing and allowed amateurs to get some breathtaking images, the humble webcam has brought amateur planetary observing back to life. Originally, the web-cams used were indeed humble, and were really sold to be perched on computers so that their owners could have video conversations with others similarly equipped. But the revolution started when amateurs decided to see what they could do when attached to a telescope. They had found a secret weapon, one that was able to beat the planetary observer's greatest foe – atmospheric *seeing*.

Take a close-up snap of the Moon or a planet with even a digital camera and you may discover that the result is often unaccountably blurred. The reason is that the seeing is distorting the fine details all the time. In the past, it was always the case that visual observers could see finer detail than any film camera, because they could pick out brief instants of steadiness that usually eluded the camera. Digital cameras fared little better, the only difference being that you could shoot away for hours for no real cost, whereas with film you were limited to 36 or so shots at a time.

Webcams are light and tiny, and have small chip and image sizes, but their great advantage is that they produce a video stream rather than single images. Used on a telescope they give a nice view of a bright object such as the Moon or a planet, jiggling about as the seeing varies, which you can record as a video file in a standard format such as AVI. But the really clever thing was the realization that some of these individual frames are of excellent quality, when the seeing steadies, so what was needed was software that would select the best ones and stack them automatically. Dutch amateur Cor Berrevoets

◀ Fig. 6.24 Bob Garner has converted his garage by a major road into an observatory for his 355 mm reflector. Here, he sets the telescope to autoguide on deep-sky objects with total exposures of many hours, in subframes of about 5 minutes. Bob's location is shown on the ISS photo on page 174.

▲ Fig. 6.25 Bob's photo of the 10th-magnitude galaxy IC 356 in the constellation Camelopardalis, taken using a total exposure time of 60 minutes through a CLS filter. To the upper right of IC 356 is another galaxy, 16th-magnitude PGC 166486, about 200 million light years away.

▶ Fig. 6.26 This is Bob Garner's photo of 12th-magnitude NGC 2336 in Camelopardalis. Below it is 17th-magnitude galaxy PGC 213387.

▼ Fig. 6.27 The historic photo of the Orion Nebula taken from Ealing in west London by Andrew Ainslie Common in 1883 using a 36-inch reflector. He prepared his own glass plates and used an exposure time of 37 minutes. For his work he was awarded the Gold Medal of the Royal Astronomical Society.

produced the original version of his free Registax software to do this in 2002, and its latest version remains one of the most useful tools available to amateur astronomers.

The camera that transformed the scene was the Philips ToUcam Pro, which originally cost about the same as the cheapest digital cameras when it was introduced in 2002 (Figure 6.28). It has a color CCD sensor rather than the less sensitive CMOS chips which are used today, has a lens that is easily unscrewed, and has easy-to-use software that allows good control of the exposure and frame rate. The combination of the ToUcam Pro and Registax started to produce a flow of planetary images that were streets ahead of anything taken using film, or even with far more costly astronomical CCDs.

The secret lies in the vast production of images. At a frame rate of 15 per second, a ToUcam will produce 1,000 images in just over a minute. Modern cameras using upgraded connections are faster still. Even if only a small fraction of these are good, they still provide enough images that can be stacked to reduce the noise. And through the magic of what is called *wavelet processing*, the resulting images can be sharpened up to reveal detail that even the keenest eye could not have spotted in the individual images (Figure 6.29).

With a webcam and Registax, amateurs started to produce images of the bright planets, the Moon and the Sun of incredible quality. Up to that time, the received wisdom was that the best instrument for planetary observing was a 150 mm (6-inch) apochromatic refractor, because it offered good contrast and typical seeing would not often make larger apertures worthwhile. But with webcam imaging, people found that they could use larger reflectors, and even SCTs with their reduced fine contrast, to get good images. Location played less of a role, as well. Suburban amateurs in their gardens can now produce images of excellent quality when the seeing allows (Figure 6.30).

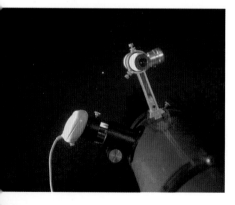

◀ Fig. 6.28 The original Philips ToUcam being used on a low-cost Sky-Watcher 130 reflector to photograph Jupiter. The unit (minus its stand, which is removed here) weighs only 60 g (2 oz), so it can be used with virtually any telescope without balance problems. It produces a video stream of 640 × 480 pixels, which is still standard with webcam-type cameras.

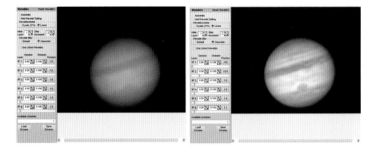

▲ *Fig. 6.29 Part of the Registax wavelet-processing screen. At left, a stacked image of Jupiter with a 200 mm Meade LX90 telescope, showing the sort of detail that you would see at first glance through the eyepiece. At right, a wavelet-processed version which reveals a large amount of detail that would require excellent eyesight and good fortune to see visually.*

Strangely, the images from webcams have more resolution than seems theoretically possible. At a mountain-top observatory, seeing better than 1 arc second is regarded as excellent, and much sought after. Yet details in amateur images made from sea level with comparatively small instruments are not only finer than this, but better than the telescope should be able to achieve. The reason lies in the way that seeing and theoretical resolution are defined. Both refer to the size of the false disk, the *Airy Disk*, that results from the diffraction of light when it passes through the telescope aperture. But with image

▶ *Fig. 6.30 This superb image of Mars was taken by Damian Peach using a Celestron C14 355 mm Schmidt-Cassegrain telescope from Loudwater, Buckinghamshire, in 2005 when Mars was high in the sky as seen from the UK and was at a favorable opposition. The bright spot is cloud covering Olympus Mons, the largest volcano in the Solar System at 27 km high.*

processing, these definitions are not appropriate. The theoretical performance of telescopes was originally estimated from visual observations using small telescopes, for example.

While many people still use their old ToUcams to good effect, these cameras are no longer available. Most modern everyday webcams are not as sensitive, and are not as easy to convert for use with a telescope by removing the lens, though a search of the web will find instructions for using a range of cameras, some so cheap that you can happily set to work on them with a soldering iron without too much worry. However, these days there is a range of purpose-built cameras to choose from. The market evolves quickly, and any suggestions I give here will be out of date next year. But typically the cameras have quite small sensors – smaller than those of DSLRs or even CCD cameras designed for deep-sky work – and quite limited resolution, often only 640 × 480 pixels (0.3 megapixels). Most smartphone users would sneer at this but it's quite large enough for planetary imaging, as planets are really quite small, though it is rather restricted for lunar and solar work. Another requirement is a fast readout rate so that the images can be transferred to the computer without having to be compressed, which reduces image quality. Webcams and their successors are the cameras of choice where seeing is the limitation rather than the faintness of the object (Figure 6.31).

Most are limited to exposures of a fraction of a second only, but some can give exposures of many minutes and will serve as deep-sky cameras as well, though they are not cooled. And as with CCD cameras, you can get both mono and color versions. A single-shot color webcam does have the advantage of being more convenient for imaging Jupiter in particular, which rotates so quickly that the features change over a matter of minutes. Those with mono cameras need to use additional software such as WinJUPOS that will take account of the difference in position of the planet between the first and last exposures.

Planetary images at the normal focus point of even SCTs with a focal length of 2,000 mm are

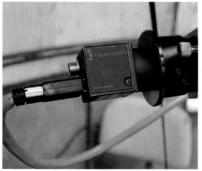

◀ Fig. 6.31 The Point Grey Flea 3 webcam-type camera is widely used for planetary and solar imaging. It connects to a computer by USB 3.0, which gives a much higher data transfer than the older USB 2.0, so it can be used at 60 frames per second.

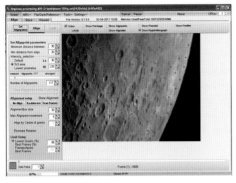

◀ Fig. 6.32 A screen grab from Registax 6.0 with the alignment points chosen automatically. The next stages are to Align, then limit the number of frames used, Stack, then use Wavelets to sharpen the image.

too small to be useful, so you always need to increase your effective focal length using a Barlow lens or similar, such as the TeleVue 5× Powermate, or eyepiece projection. If you have only a 2× Barlow, one tip is that you can increase its power simply by using a plain extension tube between the Barlow and the camera. This can easily double the effective focal length again. If you don't have an extension tube to hand, use a star diagonal, which gives exactly the same result though with a reversed image. In this way, you can increase the focal ratio of an *f*/10 instrument to something like *f*/40, giving good-sized planetary images, though you may need to increase the gain to compensate for the dimmer but larger image.

Registax is still widely used for image processing. Like much software it offers you so many options that to start with it can be daunting. Here's a very quick get-you-going tutorial using Registax 6, though some people still prefer the simplicity of earlier versions of Registax (Figure 6.32).

Click on the Select tab at the top left and choose the video sequence to be stacked. The first frame should appear in the window. If it is a particularly poor one, look for a tiny slider at the bottom of the window, just above a green square, that you can use to move from frame to frame to find a better one. Click on Set Alignpoints and the software will choose a number of points across the image and mark them with red dots. Now click on Align and it will align all the frames, allowing for any movement between them. There is a progress bar at bottom left to reassure you that something is happening.

When this reaches 100%, the software will have put the frames in order of quality and you should see a registration graph with a red line that shows you what you have. It will have suggested a cutoff limit indicated by a vertical green bar, and you need to limit the number of frames that you want to use by using the aforementioned slider, which now moves the limit bar. People tend to choose more rather than fewer, but you can experiment. Once you've done this, click on Limit and Registax will stack the chosen images. Click on the Wavelet tab when this is all done. The image may not yet look significantly better, but it should have less noise than any of the original frames.

There are several sliders to play with. The top ones sharpen the fine detail, and the bottom ones the general detail. You could be amazed by how much detail comes into view! Use them cautiously, and maybe stick to the second one down to start with. Beginners often over-sharpen, which gives dark lines around features and worsens the noise level. You will see changes as you move the slider, though only over a part of the image. Once you are happy, click on Do All to process the whole image, and save it to your computer.

When you have got all this working, you can start to experiment with different settings. Notice that you can save and load wavelet settings that work well for different types of object. You can also try out other stacking software, such as Autostakkert.

Photography using video cameras

While everyday camcorders give excellent HD video image quality, most have non-interchangeable lenses so they are not particularly suitable for serious astro-imaging. But there are some small and sensitive video cameras with removable lenses that are useful for astronomical work as their exposure time can be varied over a wide range, up to many seconds. The main practical difference between these and long-exposure webcams is that they provide a composite video signal suitable for a TV rather than a video file that you save to the computer. Many people use them with their laptops with additional framestores, however, and they can give images of deep-sky objects within seconds from suburban locations when attached to a telescope. Saving the images is not as easy as with a CCD camera (Figure 6.33).

▶ Fig. 6.33 The galaxy M101, imaged using a Watec 120N video camera. This was a stack of eight 8-second exposures made through an 80 mm refractor in a light-polluted sky. A single exposure showed similar detail but with less contrast.

They can also be used in real time for detecting meteors or aurorae and as video finders as they will give a live feed of stars much fainter than you can see by eye, even from suburban locations

Photographing different types of object

The Sun

Solar photography suffers from the same constraints as solar observing. Just because you are looking through the viewfinder of a camera, do not imagine that the Sun's intensity will not harm your eyes. You must adopt the same precautions as for visual observing. And, as with visual work, the safest procedure is to use a solar filter over the front of the telescope. You can use the brighter photographic version if you are using a webcam, or are always extending the focal length which will give a dimmer image.

DSLRs are good for capturing whole-disk images, but the best results come from webcams as the big limitation is the seeing. Lack of light is hardly a problem! Webcams can give you superb high-resolution views of sunspots, even using quite small apertures – 100 mm is standard – and solar granulation is now recorded as a matter of course whereas with film it was only hinted at (Figure 6.34). This rice-grain structure, as it is sometimes called, is the result of convection cells bringing energy up from below, and while it appears motionless, time-lapse images show that it is constantly roiling.

Needless to say, there is no threat from light pollution, though by day the seeing could be much worse than at night, given that concrete and asphalt heat up quickly. You may also find that your seeing changes as the line of sight to the Sun varies during the course of the day.

Where webcams fail is in giving high-resolution whole-disk images in a single shot. Mosaics are possible, but unless there are an unusually large number of sunspots, the general lack of detail in the solar image makes it hard to combine the images.

The same techniques apply whether you are taking images in white light or narrowband,

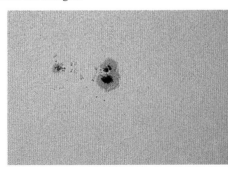

▶ Fig. 6.34 Detail of a sunspot and solar granulation photographed by David Arditti from Edgware in London through a 127 mm Schmidt-Cassegrain telescope with a DMK camera. He used an infrared filter as the seeing is generally better at longer wavelengths.

usually hydrogen-alpha. But in this case, much less exposure is needed for getting good detail on the disk compared with the prominences at the limb, and most images are combinations of two separate shots. Again, you need a good working knowledge of image-processing software which allows you to superimpose two images as layers and then combine them.

The Moon

The Moon's size and brightness make it by far the easiest Solar System body to photograph, and of course light pollution is no problem. Anybody can get a passable picture of the Moon, but I have seen some pretty poor pictures of it taken with apertures of 250 mm (10 inches), and some good ones taken with telescopes as small as 80 mm (3 inches).

As for exposure times, it is often said that the Moon should be treated exactly the same as any other sunlit landscape. It does not matter that it is a quarter of a million miles away, just as it does not matter that some parts of a landscape on Earth are a quarter of a mile away – they look just as bright as nearby objects. In fact, if you think of those occasions when you can see the Moon in the afternoon sky, it is obvious that an ordinary daytime exposure will work.

Although the Moon is a sunlit landscape, its surface rocks are largely dark basalt, so it does need a little more exposure than a view of the local park. But in general, if you can take a picture at ISO 100 with an exposure of $\frac{1}{125}$ second at $f/11$ on a sunny day, then the same is true for the full Moon. So with an $f/10$ SCT, a good lunar shot would need

an exposure of $\frac{1}{125}$ second. This is true only when the Moon is high in the sky, so allow a longer exposure if it is lower or if the sky is not perfectly transparent. With an $f/6$ telescope, give about a third as much (that is, six squared divided by ten squared to get the proportion of the areas), or an exposure of $\frac{1}{500}$ second. The first-

◀ Fig. 6.35 The Mare Nectaris region of the Moon, with the prominent craters Theophilus, Cyrillus and Catharina and the Altai Scarp, photographed by Ian Sharp using a 280 mm C11 Schmidt-Cassegrain telescope.

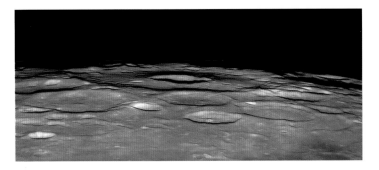

▲ *Fig. 6.36 The crater Einstein lies close to the lunar limb, and can only be seen at favorable occasions of the Moon's tilt or libration. This view by Damian Peach was taken very close to full Moon using a C9.25 230 mm Schmidt-Cassegrain telescope.*

quarter or crescent Moon need considerably more exposure because the Sun is no longer directly behind you and the landscapes are partly in shadow (Figure 6.35).

With exposure times of a fraction of a second, the telescope does not even need to be driven. You should be able to take some perfectly satisfactory pictures of the full Moon in this way. But for detailed shots, webcams rule. Even at prime focus most webcams show only a small part of the lunar surface, and you may not need any further amplification to the focal length to get close-ups. Most basic telescope drives are adequate for this, as well – even the simplest will hold an area of the Moon in the field of view for a minute or so while you take a sequence of a several hundred frames. In fact, even an undriven telescope can give you a few dozen frames of part of the lunar surface which, when stacked, will give you a nice image. You'll find that a mosaic of webcam shots will usually far exceed the resolution of a single shot with a DSLR. But you need to use the same exposure across the whole lunar disk, and also the same color balance, or you will be faced with great difficulties when you come to stitch the images together. Autostitching software is available, as it's widely used for everyday photography, and will to some extent compensate for brightness differences.

One thing that may surprise you is that despite your best efforts to choose your individual frames across the lunar surface, you end up missing a bit. This is often because the Moon can change its declination quite quickly, so if you use a pattern which moves from side to side across the Moon it may have an additional movement of its own, frustrating your efforts.

Although the first-quarter Moon is visually the most spectacular, don't give up completely on the later phases when most of the Moon's surface is just a mottled mass. Look at the limb regions where there is still some shadow detail. Remember that exposure times are less at full Moon, so you can get dramatic oblique shots more easily (Figure 6.36).

The planets

Once you have mastered lunar photography at higher magnification you can proceed to the brighter planets. While you can get good results at prime focus on Jupiter, experienced planetary imagers say that it's best to aim for the largest image scale that you can without losing too much light. There are calculations that take into account image sampling theory, pixel size, theoretical resolution and so on, but they are often proved wrong by individual circumstances. The great thing about digital imaging is that you can try different options for yourself.

Jupiter is usually the easiest planet to image because it is large and contains a wealth of detail that changes quite quickly, because of both the planet's 10-hour rotation period and real changes over a period of days and weeks. You can see the effects of the planet's rotation in

▲ *Fig. 6.37 Dave Tyler took this image of Jupiter and Ganymede from his home observatory along my road, using a C14 (355 mm) Celestron Schmidt-Cassegrain* *telescope, the instrument of choice among many planetary observers these days. The inset shows an enlargement of the image of Ganymede.*

▶ *Fig. 6.38 This beautiful image of Saturn at opposition in 2005 was taken jointly by Damian Peach and Dave Tyler using 355 mm and 230 mm telescopes from separate locations within 1 km of my home. They combined the results in image processing. The rings are particularly bright for a week or so on either side of opposition as a result of their ice particles acting like reflective road studs. Ring B is brilliant white.*

only a few minutes of observing, so you do need to limit your video sequences to not much longer than a minute. With small instruments the Galilean satellites may not be visible on frames that are well exposed for the planet, so you may need to combine images to show them as well as surface detail. Observers with large instruments can take amazing images showing detail on the satellites themselves (Figure 6.37).

Saturn is of course a favorite target, and like Jupiter is in the sky every year. It requires more exposure than Jupiter, being more distant, though with the rings it is not much smaller in apparent size. Though it rotates almost as quickly, it has much less detail so you can usually afford to take more frames.

The color balance of Saturn images sometimes causes problems. But fortunately Ring B, the brightest ring, is very close to neutral in color so if you have a good enough image, use the neutral gray eyedropper of your image-editing software to balance the rest of the image (Figure 6.38). Sometimes you see images not just of Saturn but of other planets which have a strong color cast or bias. I think these are often due to people placing too much faith in the color rendering of their computer monitor. Laptop monitors in particular seem to have a tendency to have color casts which our eyes just don't notice. So using Saturn's B ring as a neutral reference can be very useful.

Oppositions of Mars take place every two years, and the planet spends much of its time as a small and distant dot. You get only a few weeks of good observing time on the planet every opposition, and even then it is usually less than half the size of Jupiter. But when it is close its surface brightness is quite high, so you can afford to push the magnification up. Bear in mind that one hemisphere is Mars is rather bland, so if at first the disk looks blank, don't give up (Figure 6.39).

◄ *Fig. 6.39 Mars in 2007 when the tilt of the planet meant that one side could appear very featureless at times, even when photographed with a 355 mm telescope, as here. Compare this with Fig. 6.30 on page 153, where the tilt of the planet in November 2005 was different so the same hemisphere had more detail.*

Webcam images of Mars in particular tend to suffer from having a darker arc inside one limb, sometimes referred to as having a rind. This tends to occur at the sharpest limb of the planet. Away from opposition, Mars can display a noticeable phase, so one limb is often softer than the other. The exact cause of this rind and methods of eliminating it are not clear, but it seems to be caused by a combination of the seeing, diffraction in the telescope and the processing.

Images of Venus can also show this rind, though maybe because the planet is usually much brighter than Mars it is not so much of a problem. With Venus you may be able to bring out the faint dark cloud patterns that are so difficult to see visually, particularly using ultraviolet filters if you use a mono camera.

Mercury is usually so small that it is a challenge to detect anything other than its phase. Its elliptical orbit means that its western elongations take it a few degrees farther from the Sun than its eastern ones. This means that it is better visible in the morning sky from the northern hemisphere or the evening sky from the southern hemisphere. But with good seeing markings can be detected, so don't give up on it.

Other targets include Uranus and Neptune, and even the major asteroids if you have a large enough aperture. You can image the disks of the outer planets with quite a small instrument, but seeing detail on them is another matter.

Astrometry and photometry

Most people really want to be able to take spectacular images, but the rather mundane business of measuring positions and brightnesses can be of real scientific value. Variable stars and newly discovered asteroids and comets are the objects which present themselves, and in many cases good technique is more important than having dark skies. It's not a subject you can dabble in, but it would suit someone who actually wants to make a contribution to science.

The main requirements for equipment are a telescope with a long focal length and an accurate drive or guiding system, ideally permanently mounted, and a CCD large enough to cover a fairly wide field of view. You would be joining a select band of amateurs who help professionals in their work, and if your work proves acceptable you could see your name appearing on the announcements sent out by the International Astronomical Union giving details of the latest discoveries.

Deep-sky objects and comets

Sadly it is true that the worse your light pollution, the more limited the choice of objects that you can conveniently get good results from. Out in the country you can get glorious wide-angle views of the Milky Way – but light pollution often starts to creep in around the horizon, even from what are called Dark Sky Parks. But from city sites, you do need to choose your targets more carefully.

Probably the easiest targets from urban skies are star clusters, followed by small and bright planetary nebulae and galaxies. Bright nebulae are perfectly within your range if you can use narrowband filters, but the problems come with the larger objects where the gradient of the light pollution is difficult to remove. Comets have most of the same characteristics of deep-sky objects with the major difference that they move. Even quite a remote comet will move significantly over a matter of a few minutes, and over the sort of time that elapses over several subexposures the movement is really obvious. So when stacking the images you need to stack on the comet rather than the stars, which will show trails (Figure 6.40).

▶ *Fig. 6.40 Comet 103P Hartley photographed on November 15, 2010, using an 80 mm ED refractor with a Canon 40D DSLR at ISO 1600. This is a combination of 13 frames with exposure times ranging from 20 to 60 seconds made over a period of 13 minutes. The frames were aligned on the comet with the result that the stars have trailed, and the trails vary in brightness.*

Worse still, if you have separate red, green and blue exposures the star trails will start off one color and finish up another. One solution is to create separate combined images for the stars and the comet and merge the two, removing all traces of the comet or stars as necessary, which is time-consuming.

Cooled CCDs are by far the best cameras to use for deep-sky and comet imaging from the city. I have seen amazing results even from unmodified DSLRs from dark-sky sites, but in the presence of light pollution you need all the technological help you can get, including the use of light-pollution filters, as described next.

Countering light pollution with deep-sky filters

It is not often that a miraculous cure is invented, but some of the filters now available for deep-sky work could come into that category. They do indeed work wonders – but only in the right circumstances and on the right objects. Regrettably, they are not the complete answer to the light-polluted amateur's prayer. You could spend a considerable sum of money and still find that you cannot see or photograph what you were hoping to.

The filters are designed to screw into the barrel of standard sizes of eyepiece, or into a cell specially provided on SCTs. Their aim is to cut out radiation from those parts of the spectrum where the streetlights produce their light. The simplest example, and these days the least useful, is the *didymium filter*. This is a glass filter which absorbs light at exactly the same wavelength as that emitted by a low-pressure sodium streetlight. Its effect is truly remarkable. When viewed with white light the filter looks fairly clear – it has a noticeable gray-blue color, but if you had sunglasses made of didymium glass you would say they were tinted rather than dark. Look at a sodium streetlight, however, and its light all but disappears. It has a high absorption at the main wavelength emitted by a sodium streetlight, but transmits virtually every other wavelength.

In a perfect world, all outdoor lights would be low-pressure sodium and all astronomers would simply equip themselves with didymium filters at the focus of their telescopes. At one time, professional astronomers hoped to persuade the authorities in cities near observatories to use only low-pressure sodium streetlighting, but this hope has long since disappeared.

My own experience with didymium filters is not very encouraging. I obtained one in the early 1980s, when the majority of streetlights in my area were still low-pressure sodium. While it performed impressively on individual streetlights, I found that even on a bright object such as the Orion Nebula the view was not improved by using one:

although the sky background was darkened, so was the nebula. Using my camera's exposure meter, I found that the filter required an increase in exposure of one stop – that is, it had the effect of halving the brightness and turning my 220 mm (8½-inch) telescope into a 150 mm (6-inch) one. Now, a view of the Orion Nebula through a 150 mm telescope under good conditions is impressive, but I was less than impressed with my view. I can only conclude that, even then, there were a significant number of non-sodium lights around, the effects of which were not removed by the didymium filter. Today, with a higher proportion of high-pressure sodium lighting, plus other types of lighting, the filter is of little use. Photographically, I found that pictures taken through the filter were no better than those without it. But under the right circumstances such filters could be very useful – if, for example, the only nearby illumination were low-pressure sodium.

Interference filters

The didymium filter works in the same way as an ordinary color filter – that is, by absorbing a certain part of the spectrum. But the filters that have made the most difference to observers work in a different way: they use the interference between different wavelengths of light to cancel out specific wavelengths. The precise mechanism is given in detail in physics textbooks, but basically it is very similar to the way that oil on a wet road reflects certain colors, creating the familiar rainbow appearance. These filters, sometimes called *multi-layer filters*, have layers coated as thin as the wavelength of light, and can be made to transmit chosen wavelengths.

Interference filters are immediately obvious from their appearance. Whereas filters that absorb light look the same color no matter which way you look at them, interference filters are a different color depending on whether you look through them or at light reflected from them (Figure 6.41). The colors that are not transmitted are reflected away. So a filter that transmits blue light, for example, reflects the remainder, which is yellow. What's more, the exact

▶ *Fig. 6.41 The Astronomik CLS filter is an interference filter, which reflects the colors that it does not transmit. Here you can see that the light that passes through the filter is blue, while the color reflected is magenta, with some green at extreme angles. This filter is designed for use with 2-inch eyepieces.*

THE URBAN ASTRONOMY GUIDE

wavelengths transmitted or reflected depend on the angle at which they are viewed. They are designed to be used with the light passing through at right angles to the surface, and thus produce the best results when the object being viewed is exactly on-axis, at the center of the field of view.

If your field of view is narrow, as at high power, this requirement is usually met. But with wider fields of view more light arrives at an angle. This results in stars near the edge of the field of view appearing a different color from those near the center, which can be disconcerting. When used with wide-field designs of eyepiece, such as Naglers, multilayer filters will give their best results only at the center of the field of view.

Deep-sky filters, which are also generally known as *light-pollution-reduction (LPR) filters*, fall into two broad categories. Some cut out wide bands of the spectrum in which the main light-pollution lines are to be found, and therefore transmit quite a large amount of light from objects with continuous spectra, such as stars and galaxies. Examples include Lumicon's Deep Sky filter and Orion's SkyGlow filter. Others, which are more expensive, cut out the whole spectrum except for a band which is chosen to include a particular set of emission lines from nebulae. These are called *nebular filters* or *ultra-high-contrast (UHC) filters*. Within this group there are narrowband filters which isolate specific lines.

Broadband filters

The lowest-priced multilayer filters block most green and yellow parts of the spectrum. There is one very unpleasant fact of life where filters are concerned. We have evolved on a planet orbiting a completely average star, and our eyes are naturally most sensitive to the peak of its radiation, which in bright conditions is yellow light. This is why the artificial light we tend to prefer, for streetlighting and other purposes, is predominantly either yellow or white. Since other galaxies and clusters consist, on the whole, of other average stars, their light is roughly the same color as many of our streetlights. So by cutting out the light from a streetlight, particularly a high-pressure sodium one, we are also cutting out a major part of the light of a distant star cluster or galaxy.

This means that no light-pollution filter can be totally satisfactory for viewing the majority of deep-sky objects, which are either galaxies or clusters. If you are fortunate enough to live where most of the streetlights are blocked out by simple LPR filters, the contrast of deep-sky objects against the background can be improved. The effect will be to reduce the aperture of your telescope, so you should be prepared to take every possible step to increase contrast rather than relying on the filter to wipe out light pollution.

You can read many reviews and user comments of broadband filters by doing a simple web search, but many of them may be contradictory. One reason for this is that not everyone's light pollution is the same, even within a fairly small area. As we saw in Chapter 3, there are many different light sources around, each with its own color distribution. Someone who lives in an area where there are many mercury or LED lamps will have different light pollution from one where there are low-pressure sodium lamps. But in general, broadband filters have very limited value in the presence of strong light pollution. The difference they make can be so minor as to be undetectable. In a less badly light-polluted area, however, they can have enough effect to make the difference between seeing an object and not seeing it. Being told that "it might work or it might not" does sound like rather pointless advice, but the fact is that you do have to try them for yourself in your own environment. This is where being a member of a local astronomical society, or at least getting to know another local astronomer, can be a good idea.

Nebular filters

Undoubtedly the most useful multilayer filters are those which cut out virtually all light except for a relatively small band that includes a few emission lines. Examples include UHC filters from various manufacturers, Astronomik CLS filters, and, with even narrower bandwidths, O III and H-beta filters. These make use of the fact that most gaseous nebulae (with the exception of reflection nebulae, which appear blue on photographs) emit bright-line rather than continuous spectra. The O III (pronounced "O-three") filter, for example, allows through only the green lines of doubly ionized oxygen at 496 and 501 nm. By viewing them in the light of these lines only, you can in theory observe the light from the nebula, and cut out the rest (see Figure 6.42). This removes almost completely the light from low-pressure sodium and mercury, and a large part of tungsten and fluorescent. High-pressure sodium has a line very close to the important O III lines, so it is not reduced as much as one would like. Fortunately, the peak of human night-vision color sensitivity is close to the O III lines.

A typical UHC nebular filter for visual observing has a bandpass (width of transmission band) of some 25 nm, which includes the green lines of H-beta and O III. This allows you to view most H II regions and planetary nebulae. Other filters are available that transmit an even narrower band, so that only the light of either H-beta or O III gets through.

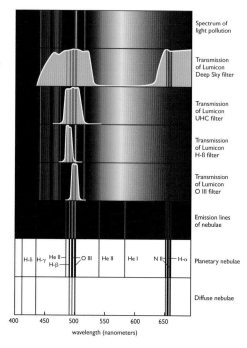

► *Fig. 6.42 Transmission curves of Lumicon filters, compared with a typical spectrum of light pollution (mostly high-pressure sodium, with some mercury) and important nebular lines. In each case the transmission of the filter is shown as a window which admits particular lines from nebulae while excluding most light pollution. The light-pollution line at around 500 nm, caused by high-pressure sodium, is very close to the important O III lines in the spectra of planetary nebulae.*

A filter such as a UHC is ideally suited to observing emission nebulae in whose spectra are lines of hydrogen, oxygen or nitrogen. This includes H II regions such as the Lagoon Nebula (M8) and the Omega or Swan Nebula (M17), both in Sagittarius. It is also good on most planetary nebulae, such as the Owl Nebula (M97) in Ursa Major, and on supernova remnants such as the Crab Nebula (M1) in Taurus, which can be surprisingly hard to see from urban areas on account of its fairly low surface brightness. These filters are not likely to be of much use when viewing star clusters, galaxies or the blue reflection nebulae, as they will cut down the brightness of the object in the same proportion as the light pollution.

The H-beta filter allows through only a 9 nm band centered on the green H-beta line at 485.6 nm. Oddly enough, it is best for very faint objects that appear reddish-pink in photographs, such as the elusive nebula IC 434 surrounding the Horsehead in Orion, or the California Nebula (IC 1499) in Perseus. Their pink color is a combination of H-alpha, which registers on the red channel of an image, the weaker H-beta, which registers on the green, and H-gamma, on the blue. The dark-adapted eye is most sensitive to the green H-beta line, and not at all to the red. This filter works best on objects that give out only hydrogen emission, and show virtually no other lines, and is therefore most appropriate for nebulae which are glowing under the influence of a star some distance away, so that only hydrogen, which is readily excited, is glowing. It is not particularly effective on most planetary nebulae, where the star that is causing the nebula to glow is closer and therefore able to excite oxygen and nitrogen too. Some planetaries, however, have a low-temperature central star which is not able to generate the O III lines. An example is NGC 40 in Cepheus, a magnitude 10.5 object about half the size of M76. These filters, then, have a very limited use and are probably most suitable for the dedicated deep-sky observer. It is also quite likely that from badly light-polluted skies even an H-beta filter will not show the faint nebulae. Lumicon claim that you will be able to see the Horsehead Nebula as long as the sky is dark enough for you to easily see the (northern hemisphere) winter Milky Way. An H-beta filter is often referred to as the Horsehead filter, as this is just about the only object it is good for.

For viewing most planetary nebulae an O III filter is more appropriate, as their output of this green line is stronger than their hydrogen lines. Nick Hewitt finds that the notoriously large but low-surface-brightness Helix Nebula (NGC 7293) in Aquarius is "pretty easy with an O III filter" in skies in which it is otherwise invisible. I have seen it without a filter using 10 × 30 image-stabilized binoculars from a very dark site in the west of Ireland during a Burren Star Party, but it was not an easy sight even from there. From the British Isles it is quite low in the sky, which doesn't help.

I can see the Veil Nebula (NGC 6992) in Cygnus fairly easily using an O III filter on a 130 mm reflector from my current site near High Wycombe, a town of some 120,000 people about 30 miles from London. For years the Veil was regarded as one of the most challenging for the deep-sky observer, even from dark sites, but it is now visible with little difficulty from all but the worst locations, despite the great increase in streetlighting, thanks to ultra-high-contrast filters (see Figure 6.42). They are the nearest thing we have to "magic filters" that will just cut through the light pollution. However, I would not recommend narrowband filters for use with telescopes significantly smaller than 130 mm (5 inches). As it is, you need to keep your eyes well dark-adapted and ideally put a black cloth over your head to keep out extraneous light.

Owen Brazell of the Webb Society and the BAA's Deep Sky Section has considerable experience of using filters from a variety of sites in Britain and North America. He feels that all but the hard core of deep-sky observers will find the nebular filters such as the UHC more suitable than narrowband filters. "Because the filter passband is relatively wide, the attenuation of starlight is not too great and the

▶ Fig. 6.43 Nick Hart lives in the city
of Newport, south Wales, which has a
major port and steel works. As you can
see from the general night view, light
pollution is a problem. Nick has an
observatory housing a 250 mm f/4.8
reflector with which he takes photos
using an Atik 383L mono camera and
2-inch Astronomik filters in a Starlight
Xpress filter wheel. He also uses a
100 mm (4-inch) f/7 refractor with
focal reducer making it f/6.3. His images
show that you can take deep-sky images
in the presence of light pollution if you
choose your targets and filters carefully.

▼ Fig. 6.44 This H II region in
Cassiopeia, NGC 281, is also known
as the PacMan Nebula because of
its shape, remniniscent of the 1980s
computer game icon. This image from
Nick shows the standard depiction, with
40-minute exposures in R, G and B using
the 250 mm reflector.

▲ Fig. 6.45 In this version, Nick
used additional exposures of 60
minutes each using H-alpha and
O III filters which were added
to the red and green exposures
respectively, giving a color
rendering which differs from the
usual version. Sometimes an S II
filter is used for the red frame,
with the H-alpha being used for
green and the O III for the blue,
giving a very different rendering.

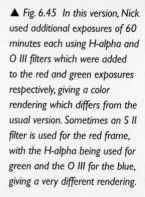

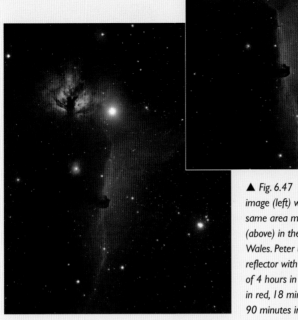

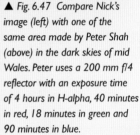

▲ Fig. 6.47 Compare Nick's image (left) with one of the same area made by Peter Shah (above) in the dark skies of mid Wales. Peter uses a 200 mm f/4 reflector with an exposure time of 4 hours in H-alpha, 40 minutes in red, 18 minutes in green and 90 minutes in blue.

▲ Fig. 6.46 Nick's image of the Horsehead and Flame nebulae in Orion was made with the 100 mm f/6.3 refractor. He gave 60 minutes through each R, G and B filters and 3 hours through a Lumicon Night Sky H-alpha filter. He used 10-minute subframes. The same area can be seen on short exposures made with film and a DSLR in Fig. 4.28 on page 98.

▼ Fig. 6.48 Nick's image of M51 was taken with 60-minute total exposures in each of R, G and B filters, in 5-minute subframes.

views of nebulosity associated with stars (such as M8 and M16) are very pleasing," he reports. They will help cut down the sky background even from heavily light-polluted neighborhoods and allow you to see objects that would otherwise have been at the threshold of vision. He warns, however, that the narrowband filters have very restricted uses:

> "The view of M42, for instance, through an O III filter is, quite frankly, disappointing. But if your interest is in hunting out faint and obscure planetary nebulae then this filter is a must. These objects will simply be invisible without it. The detail which can be brought out in nebulosity with this filter is incredible."

Star images are dimmed by about five magnitudes, so some of the aesthetic appeal of a nebula in a good starfield is lost. The O III filter works on most planetaries, but not on the Crab Nebula, whose light is produced by a different mechanism – *synchrotron radiation.*

Owen also reports that you can make interesting comparisons of the views of bright objects such as the Orion Nebula through O III and H-beta filters. The O III filter shows detail only in the central regions of the nebula, where the ionization is strongest, while the H-beta will reveal detail in the outer parts that cannot be discerned by any other method. David Cortner, observing from Tennessee, says that on bright nebulae:

> "A good LPR filter (like Lumicon's UHC or Orion's UltraBlock) can help salvage most of what streetlights take away. M16, M17, M8, and M20 are all helped immeasurably by the UltraBlock filter; M42 doesn't need the filter's help; and fainter nebulae walk a fine line of invisibility. The filter increases contrast but at some cost in brightness – if the latter effect takes faint nebulae below the limit, the increase in contrast is worthless. Faint nebulae in Auriga and Cassiopeia fall into this category. Lesser nebulae in Orion (east of Zeta, north of M42) are sometimes helped, sometimes hindered by the filter, depending on sky transparency and brightness."

As for planetary nebulae, David's comment is that:

> "What filters can do for bright nebulae, they can do for planetary nebulae in spades. Besides, some planetaries have such extraor-dinary surface brightness that they shine through the glowing sky without ill effect. These also bear up well under high magnifica-tion (which, in turn, darkens the sky 'foreground'). The Saturn Nebula, M57, M27, the Eskimo, all shine nicely for me. The

Helix is visible on the very best nights from here, despite its low elevation and its position over the worst of my city's glow. It is huge, of course, and to see it at all I need the LPR and the 16-inch (400 mm) Dob working with a wide-angle eyepiece. The Veil can be beautiful when near the zenith and with the LPR filter. It's easy to see in the Dob with the filter and often shows good filamentary structure – both arcs are usually easy, and on the best nights I can even see the brightest knots in the middle. Only the brightest portions of the two bright arcs appear without the filter, and then without any detail. In the [125 mm/5-inch] refractor, the Veil is utterly invisible without the filter from here, and only dimly so with it. But as is so often the case, the very wide view in the refractor and the diamond-hard stars add a 'context' to the Veil which makes up for much of the brightness lost in the smaller glass."

Nebular filters are still worth taking with you on trips to dark sites. From the caldera on Tenerife in the Canary Islands, at an altitude of around 2,000 meters (7,000 feet), I had a good view of the Veil through a 220 mm (8½-inch) reflector equipped with a Lumicon UHC filter. Without the filter, the Veil was considerably less easy to see. As mentioned in Chapter 7, the skies of Tenerife, while excellent by city standards, do suffer from the lights of coastal resorts. In addition, there is natural skyglow, which means that the sky between the stars is never completely black.

For comet observers, Lumicon offer their Swan Band Filter. It transmits a 25 nm band which includes the green lines due to carbon at 514 nm emitted by some comets, notably those which have a high gas content rather than dust. This is fine for visual use on some comets, but it does not work for them all, and nor is it suitable for photographing comets.

Photography with deep-sky filters

Deep-sky filters are usually available only in eyepiece sizes, though as this includes 2-inch barrels it is often possible to use them in front of photographic lenses, though insecurely. However, a better solution is to use a clip filter, available for most Canon DSLR cameras from Astronomik. This fits between the lens and the sensor, with the advantages that it is much cheaper than buying large filters for each lens, and will work with virtually any lens, even extreme wide-angle ones, as the light passes through it more or less at right angles to the filter, so there are no off-axis rays to cause funny background colors (Figure 6.53).

▲ Fig. 6.49 International Space Station photo of the British Isles at night, with the locations of four astrophotographers indicated by circles. A: Nick Hart, Newport, south Wales; B: Peter Shah, Meifod, mid Wales; C: Bob Garner, Greenford, west London; D: Bob Winter, Archway, north London.

▼ Fig. 6.50 Galaxies are particularly challenging from Bob Winter's home site in north London, and for this version of M33 he used his Takahashi FSQ85 and QSI 532 camera with luminance, H-alpha, R, G and B filters for a total exposure of 90 minutes.

▲ Fig. 6.51 The Cone Nebula in Monoceros is at the right of this view taken by Bob Winter from his north London rooftop. He used a Takahashi FSQ85 85 mm f/5.3 ED refractor and a QSI 532 camera with H-alpha and O III narrowband filters plus a luminance exposure, with a total exposure time of 90 minutes. The designation NGC 2264 refers to both the Cone Nebula and the star cluster which occupies a wider area of the picture, known from its shape as the Christmas Tree Cluster.

▶ Fig. 6.52 The Little Dumbbell Nebula in Perseus is a planetary nebula so it is particularly suited to city imaging. Bob Winter used a Takahashi TSA102 102 mm f/8 ED refractor with a Starlight Xpress SXVR-H9 mono CCD camera. His luminance, H-alpha and O III exposures totaled 100 minutes.

You can get a wide range of different filters for photographic use, including O III, H-alpha and S II (sulfur) at 672 nm. The photographic versions are made with narrower bandwidths than the visual ones, with the great advantage that they cut out virtually all light pollution. So even from the city it is possible to take images of faint gaseous nebulae, given long enough exposures and sufficient subframes.

Bob Winter lives in Archway, north London, where the stars are hardly visible, yet he manages to image nebulae in color. He uses a CCD and narrowband filters and gets results that would have been impressive in the last century using film even from desert areas. He comments:

"I use a monochrome, cooled astronomical camera with a filter wheel containing narrowband and broadband filters. In urban conditions I don't think single shot color cameras or DSLR cameras give worthwhile results. When selecting targets I find it's best to choose emission objects such as star-forming regions and supernova remnants. These can be imaged with narrowband filters, thereby cutting out most of the urban pollution. Broadband objects such as galaxies can still be imaged, but may need additional anti-pollution filters, such as a CLS filter. On the subject of filters, I have found that the Astrodon range, especially the narrowband ones, 5 nm H-alpha and O III, give a big improvement in image quality. Expensive but very effective. Finally, correct software is essential. There's no point in having an excellent telescope and running it with substandard software. I use Maxim DL for camera control and guiding and full Photoshop for final image processing."

◀ Fig. 6.53 At left, Auriga photographed through a CLS filter placed in front of the camera lens shows different sky background colors at the edge compared with the middle of the field of view. At right, Orion photographed with a similar lens, but using a clip filter inside the camera, shows no variation in color.

7 • IF ALL ELSE FAILS ...

After all the tips and advice given in this book so far, there is one sure-fire way to overcome the light pollution and see all those objects you have read about: go where the skies are really dark. Many amateur astronomers now choose this option, and seldom if ever observe from their home. Others divide their time, observing the most accessible objects from home and waiting until they can visit dark skies to look for nebulae and other deep-sky objects.

While it might seem a simple option, traveling to a dark site is not always easy. Some people are lucky enough to live where light pollution is limited to the urban area itself, and can get away to dark skies fairly quickly. I once enjoyed a good night's observing just 16 km (10 miles) from downtown Wagga Wagga, in Australia. The town, which is brightly lit, was just a glow on the horizon and everything else around was dark. Maybe the locals would consider that poor conditions, and would go farther out, but it is all a question of what you are used to.

Most people will have to do some traveling even to get to tolerably dark skies, in which the Milky Way is visible and there is at least a fighting chance of seeing faint objects. This probably means a drive of an hour or more, and using the car as a base – the sort of trip you can do in a single night. But for really dark conditions, a major journey is called for, either a very long drive, with overnight stops, or a plane journey. Indeed, for non-drivers every trip to a dark site takes on the flavor of an expedition. In this chapter I look at what sorts of site you might try, and discuss the equipment you need for portable astronomy.

Choosing a site

Satellite views of the Earth at night show us what we are up against (see Figure 1.2 on page 9). The builtup areas of populous countries are marked by sprawling blobs of light. Satellite photographs of London show my own location on the fringe of London to be well within a zone of brilliance some 105 km (65 miles) long, centered on London itself.

Despite the fact that I am on the edge of open countryside, there are numerous major roads and small towns in every direction, each contributing to the sky brightness. But immediately local lights do usually make a considerable difference, and just driving a couple of miles away from the builtup area brings a significant improvement. The same applies to inner-city environments. Finding a spot some distance away from the local lights could make your life easier. The problem is finding a spot where you can observe in safety and without arousing suspicion!

In general, the farther you can get from the major builtup areas the better. A big metropolis such as London or New York spreads its light over a very wide area, but smaller towns and cities are not nearly as bad and you can escape from them much more easily. Satellite photographs do help in choosing a dark site, but they do not tell the whole story. They show only areas where there are upward-shining lights.

You can see this very clearly in the ISS image of London taken in February 2013 (Figure 7.1). The M25 ring road shows up very brightly, even along areas where the lighting is full-cutoff. What you can see is the reflection from the road surface. This is a snapshot of the light emitted through a very narrow angle vertically upward. Even the suburbs with their non-cutoff lamps show only dimly, as their more numerous roads are lit to a lower road brightness. The center of the city shows up brilliantly, as a result of the floodlighting of buildings, advertising signs and so on. This view does not show the contribution to skyglow from lights that shine at other angles to the vertical, which nevertheless contribute greatly to light pollution. You would need a series of images of the same city taken as the Space Station approaches, so as to give a picture of the light emitted at different angles.

▲ Fig. 7.1 London as seen at night from the International Space Station. This viewpoint shows light that shines directly upward, with floodlights on buildings being major offenders. Roads that are lit to major road standards show up clearly due to reflection from the road surface.

▲ Fig. 7.2 Light pollution extends far from the actual lights, as is shown by this photograph of the night-time view south from Portland Bill. Although the site is on a promontory that juts out several miles from the south coast of England, it records star trails in a pale blue sky tinged with yellow, and a band of cloud along the horizon. The bright line along the horizon is the trail of a ship's lights.

Even so, there is a lot to be seen in these pictures, which are available online for many areas if you search the archive. Notice bright blobs along the M25. These are the overhead gantries carrying signs illuminated from below by six lamps. Some of the light spills upward, and can even be seen from space. Even individual floodlit buildings are visible in some cases. Major contributors to skyglow over rural towns and villages are local churches, whose floodlighting ironically helps to destroy our view of the heavens. These floodlights must really improve the security against thieves arriving by parachute!

I estimate that for most purposes you need to be about 80 km (50 miles) away from a major city for the effect of its lights to be negligible for observational purposes. Although you will still be able to see a glow in the sky, it should not be objectionable. On mainland Britain, the only places where you can be this far from all major cities are the heart of Wales and the north of Scotland. To be completely free from the effects of upward-shining lights, you would have to be about 275 km (170 miles) away from them. To get this figure I have made the arbitrary assumption that the densest part of the atmosphere is within 5 km (3 miles) of the surface; almost exactly half the atmosphere is below 5 km. An air molecule 5 km above you will have a horizon 275 km away, and so will just be illuminated by the lights, however faintly, but the air mass below it will be hidden from the lights by the shadow cast by the Earth.

The coast might seem a good place to head for to get away from light pollution, but a photograph I took at Portland Bill (see Figure 7.2), a promontory on England's south coast, shows that this is not the case. The French coast is 100 km (60 miles) away, yet the time exposure of the sky to the south shows the unmistakable sign of light pollution. The lights of the resorts along the English south coast shine out a considerable distance.

Your choice of observing site depends not just on the darkness of the skies. If you are looking for a site within reasonable driving time from your home, and plan to return home after the night's observing session, you will not want to travel for much more than an hour or two unless you can sleep in your vehicle. You will also need to find a site well away from other people, for various reasons. You do not want car headlights shining on you just as you have become dark-adapted, and furthermore, although it can be fun to show the wonders of the heavens to inquisitive strangers, they can interfere with your program of work. Setting up at the edge of a public road is rarely convenient, and in the UK at least rural parking lots near towns are often occupied by people with other things on their mind, where the presence of someone armed with binoculars and cameras would be unwelcome, not to put too fine a point on it.

Some American amateurs carry firearms as a precaution against potentially dangerous situations. To those in many other countries this is unthinkable at present, but one wonders if it will become the norm everywhere in years to come. I have encountered people with guns at night in Britain, but they were interested only in rabbits. I always hope they can tell the difference between a rabbit and an astronomer.

It was to avoid such problems that some years ago a group of amateurs in the English Midlands devised a sign, in conjunction with the local police force, which they could place near their observing site to warn people that there were astronomers about. This followed an incident in which one of them was pounced upon by an alert constable attracted by the strange nocturnal activity! Thankfully, British police do not as a rule carry guns, or the outcome might have been more serious. Several times I have been the center of police attention while observing in the country, but have found that once they know what I am doing the only problem is getting rid of them. The inhabitants of country areas can be inquisitive, and are often concerned about what might be going on in their area, even if they do not own the land. The obvious answer to these problems is to observe from private land with the agreement of the owner. Although

▶ *Fig. 7.3 Members of the West of London Astronomical Society visit this site at Llanerchindda in Wales for their annual dark-sky observing weekend. Skyglow from the Cardiff area over 60 km (40 miles) away shows up in this 8-minute exposure at ISO 1250 with an f/3.5 fisheye lens, but overhead the observing conditions are excellent.*

this may take some organization, it could make for a much more pleasant and comfortable session, especially if you can observe in the lee of outbuildings on windy nights. Some astronomers in Britain lucky enough to have good sites themselves advertise their facilities to others, and offer bed-and-breakfast accommodation along with the use of telescopes. Some areas, such as in Galloway and Exmoor, are now designated as Dark Sky Parks. Look in the ads in astronomy magazines or search online for details (Figure 7.3).

The search for dark skies

Every so often, the desire for really dark skies overwhelms us. For amateurs living in the conurbations of Britain or northern Europe there are many possible sites, but it is noticeable from light-pollution photographs taken from orbit that some countries, such as Britain and the Low Countries, are heavily light-polluted compared with their more southerly European neighbors. For British observers, the simple act of crossing the Channel to France brings darker skies, once you are away from the towns. And the farther south you go, the better the chance of good weather.

There are now several observatories for the use of amateur astronomers in France, Portugal and Spain. Typically, you can either take your own instruments and have the pleasure of observing with them in good conditions, or use those provided, though this may mean having to share time with others at the site. Another site popular with European astronomers, both professional and amateur, is the Canary Islands. These Spanish-owned islands off the west coast of Africa are a four-hour flight south from the UK, at latitude 28°, so they offer much more of the sky than can be seen from northern Europe. The main island of Tenerife has the advantage of being accessible by charter flight, with package deals on accommodation which can be much cheaper than trying to book your own, particularly as transport to the resorts is included. The coastal strip is ablaze with lights (Figure 7.4), so do not expect to be able to observe much from your hotel balcony. However, Tenerife rises steeply from the sea, and just an hour's drive from the main resorts will take you to an altitude of about 2,000 meters (7,000 feet), often above an inversion layer kept low by the cold sea current. There is plenty of room to observe from above the treeline, in Las Cañadas National Park, and even in winter the temperature at this level rarely drops below 5°C (41°F).

Despite the presence of so much lighting, the skies from the center of the island are often excellent by European standards (Figure 7.5), and a large professional observatory has been built on the ridge. The main peak of Tenerife is the volcanic Pico del Teide, at 3,718 meters (12,198 feet), but there is no point in trying to observe from its summit. When British astronomers were site testing for a Northern Hemisphere Observatory, as it was then known, an intrepid team camped near the top of Teide to see what it was like. Among the hazards were the sulfurous fumes which filled their tent. A less uncomfortable site was the nearby peak of Guajara, which was chosen by Charles Piazzi Smyth for his observations of the 1834

▲ Fig. 7.4 Tenerife is favored by many European astronomers as a relatively dark and accessible site. This view from neighboring La Palma shows how the coastal fringes are strongly illuminated, though the central caldera area of Las Cañadas can still have dark skies under the right conditions.

▲ Fig. 7.5 The Milky Way is a splendid sight from a dark site. This photograph was taken by the author from near Teide Observatory on Tenerife using a 24 mm lens at f/2.8, exposing for 6 minutes on Fujichrome 1600 film. The horizon at lower left slopes because of the angle of the equatorially mounted camera.

return of Halley's Comet, and from where he also made the first infrared observations of the sky. The remains of his site can still be seen. But there is little advantage to be gained from climbing there since Las Cañadas offers many more accessible locations. You can even stay in the Parador Hotel below Guajara. Although this is expensive by Canarian standards, it does avoid having to make the long and potentially dangerous drive down to your resort after a night's observing, and there is even a telescope available for use by guests by arrangement (Figure 7.6). From this hotel I have seen shadows cast by Venus, and at the same time noticed that the ground was shimmering faintly as the light from the thin crescent was distorted by atmospheric turbulence.

There are an increasing number of such facilities available to North American amateurs. The principal ones are in western states such as Arizona and New Mexico, but there are others in eastern states as well. A web search on "astronomy village" or "astronomy resort," plus a look through the ads in astronomy magazines, will turn up quite a few.

As well as the opportunity to observe from dark skies, star parties offer the chance to compare commercial and home-built equipment. These range from small get-togethers by local astronomical societies to gatherings of hundreds of amateurs, with trade stands and other attractions. Some are attended by representatives of the major telescope firms, eager to give their goods a high profile, with complete telescope systems occasionally given away as prizes. And other amateurs who already have the equipment will gladly give their verdicts on performance. In many cases you will be able to compare for yourself, with the owner's permission. There will undoubtedly be an impressive array of telescopic firepower available, and some proud owners are only too pleased to see a long line of prospective viewers, waiting for a brief peek at the Veil Nebula or whatever happens to be on view. The biggest instruments are the most popular, not surprisingly, but even small instruments are used, as their owners are keen to find out what their own pride and joy can really achieve.

There is an etiquette to star parties, chief among which is not to use any white lights while observing is in progress. In particular, this means not driving your car on site at night, so make sure you arrive in good time, and not expecting to be able to drive away when you get tired. Many star parties are intended for serious viewing, so if you have a family make allowances for their needs, unless they are all as dedicated as you.

In the British Isles, the largest star party is held annually at the Kelling Heath campsite in north Norfolk (Figure 7.7). Organized by Loughton Astronomical Society (from northeast London), it is

held every autumn at a new Moon period close to the equinox, with another event at the same site organized by the Norwich Astronomical Society in spring. Other regular events takes place at Kielder Forest, Northumberland, and in the Isle of Wight. Among the big regular North American star parties and telescope conventions are the Riverside Telescope Makers' Conference in California, and

◀ Fig. 7.6 Martin Lewis observes with a 60 cm (24-inch) privately owned Dobsonian telescope at the Parador Hotel in Las Cañadas on Tenerife. In the background, moonlight illuminates the peak of Guajara.

▲ *Fig. 7.7 The Equinox Sky Camp at Kelling Heath in Norfolk is the UK's leading observing event. Hundreds of* amateurs converge on the campsite to enjoy much darker skies than they can see from home.

the Texas Star Party (both in May); Stellafane in Vermont, intended mainly for serious telescope makers, and Mount Kobau Star Party in British Columbia (both in August); and the Winter Star Party in Florida (January or February). In addition, the Royal Astronomical Society of Canada organizes a star party on a different site each year. *Sky & Telescope* lists events on its website or you can search online for "star parties list." You may need to book your place well in advance, since some of the big parties have to limit attendance. The best sites at the Kelling Heath event, for example, are regularly booked up within hours of pitches becoming available.

As well as established sites such as these, international trips are often arranged to dark-sky sites such as the Australian Outback or for events such as eclipses. These are advertised well in advance in astronomy magazines, and offer you the chance to get to dark skies and to meet other astronomy fanatics. The downside of these trips is that they can remove your personal freedom to choose your own site.

I have often wondered where the best skies in the world are to be found. In my own limited experience, the southern hemisphere has better skies than the northern. This is certainly the initial impression you get, partly because the Milky Way in the southern hemisphere is brighter than the northern section, and partly because your attention is drawn to familiar constellations higher in the sky. But when comparing the skies from, say, Tenerife and Brisbane, which are at similar latitudes north and south of the Equator, the impression persists. I suspect that the lesser amount of industry and, in particular, air travel has a lot to do with it. Every aircraft produces a trail of pollutants at high altitude, which provides nuclei for condensation and hence leads to milky skies.

Professional astronomers are in the position of being able to travel to the world's best astronomical sites. Some say that the mountain sites of Chile, where there are several European and American telescopes, are the best they have experienced, while others maintain that the skies at the South African Astronomical Observatory at Sutherland, on the Great Karoo plateau, are better still.

Remote observing

For years, professional astronomers have been able to observe remotely. They can sit in their office in, say, Edinburgh on a rainy day and conduct observations being carried out at night-time atop Mauna Kea in Hawai'i. A data link enables them to see what is going on and assess the quality of the observations. Such remote observing is now popular among amateurs, using large-aperture instruments at prime sites such as New Mexico and Siding Spring, Australia. This can be thought of as an extension to CCD and computer-controlled astronomy, where the observer sits watching the image appear on a computer screen rather than at the eyepiece. Instead of the telescope being a meter or so away, it can be on the other side of the world. Naturally this sort of facility costs money, but the costs and availability of instruments are not excessive and the facilities are open to all. You generally pay for only the satisfactory exposure time you use, and can take images either in real time or according to a pre-arranged plan with a variety of instruments from small refractors with single-shot color cameras to large reflectors.

Though a single individual color image with one of the larger instruments might cost you as much as a cheap digital camera, which sounds horrifying on the face of it, compare that with the costs of buying your own telescope and mount and equipping it with a good CCD and guide system. When you bear in mind that you could be observing with a half-meter telescope in a dark sky with a CCD far better than you could afford, the costs seem reasonable.

Your car – a mobile observatory

Traveling to your observing site by car has the advantage that you can take all the gear you are likely to need for the night's observing. A car is very convenient for this, but it is hardly purpose-built. There are a few modifications you can make before setting out for the rural darkness. First, get a roll of translucent red sticky tape, and use it to cover your car's interior lights and any other lights you use during the session.

Next, pay attention to your power supplies, for both your telescope drive and any camera you are going to use. Most equipment plugs into a car's cigarette-lighter socket. Some sockets require the ignition switch to be at its first position. But you should not turn the switch so far that the ignition lights light up on the dashboard as this may continuously energize the ignition circuits, which could overheat and burn out. Telescope drives draw comparatively little current from a car battery, so in theory it is unlikely that you will drain it. However, I have been caught out on a remote road in New South Wales when the battery of my hire car packed up after half an hour on parking lights. The engine would not turn over at all. Fortunately, after an hour the battery miraculously recovered, but I have never completely trusted a battery since.

Most telescope suppliers sell power packs which are basically the same as the ones designed to start your car when the battery fails. Alternatively, you can use an ordinary car battery – it need not be a new one. Just keep the old battery when you get a replacement. But bear in mind that it should be fixed in place in some convenient location, and make absolutely sure that nothing can drop across the terminals and short them out. Using a separate battery means that you can do away with that maddening cigarette-lighter connection. It is strange that just about the only universal fitment on a car is also the most infuriating. Although it may do the job of lighting cigarettes well enough, the problems start when you try to use it as a general connection for all the accessories that are now available. It is not permanent enough for my liking, and the plug can easily be pulled out. One of many frustrations that awaits owners of Go To mounts is that your careful setup procedure is completely ruined by a momentary loss of power. Instead of the drive just pausing for an instant, it loses all its knowledge so you usually have to start again from scratch and waste valuable time. Furthermore, in addition to the telescope drive you might wish to use a number of accessories that require electric power: an extra light, a portable hair dryer to get rid of dew on the camera lenses, a computer, maybe even a soldering iron to fix the connections that always come undone as you observe.

It would be best to have a different system for low-voltage power supplies, with a bank of sockets conveniently located at the back of your car so that you can simply plug in whatever you want. I do not know of any particularly good connection system. Power connections for small accessories these days are often via 3.5 mm jacks, but they are just as bad in their own way. Apart from the fact that they are very difficult to wire up, for some perplexing reason manufacturers have reached no agreement on whether the central pin should be positive or negative. Vixen mounts have a negative central pin, the opposite of most other appliances which use identical plugs. The temptation to use ordinary household plugs for low voltage should be resisted, as there is then the danger that someone – maybe even yourself – will plug your low-voltage appliance into the household supply at some stage. It looks as if we are stuck with the cigarette-lighter system, with all its faults. It is a good idea to make up a socket bank wired directly into either the car battery or a separate battery, with a device for preventing the plugs from being pulled out. One simple method is to cut a narrow slot in a piece of wood fixed to the bank of sockets. Tie knots in the cables and slip them into the slot so that if the cable is pulled, the knot prevents the plug from being pulled out.

Yet another bit of universal bad design among telescope manufacturers is to supply black cables with everything, which are guaranteed to be tripped over in the dark. If you are reorganizing your wiring, choose white cables rather than black ones.

For safety's sake it is best to stick to low-voltage devices. But should you need a higher voltage to run your telescope or even your computer, you can do so via a device called an *inverter*. These are sold for use on caravans and yachts, and provide fairly low power at domestic-supply voltage from a car battery. They are similar to the variable-frequency oscillators used to control synchronous motor drives, but they operate only at domestic-supply frequency. Bear in mind that the more powerful the unit, the greater the temptation to plug in more and more appliances, and the greater the risk of draining the battery. I have also had power adapters ruined by a basic inverter that produced a square-wave output rather than the normal household AC sine wave, so it could be worth spending more on a sine-wave-output inverter. Always use a safety device generally known as an RCD in the UK and Australia or GFCI in the US on any socket supplying AC power.

In addition to your observing equipment, it is a very good idea to take a small portable table, such as a card table. This will take your charts, observing book, pens, and of course eyepieces, filters, and so on. You may like to rig up a canopy over it to protect everything from dew or frost, together with a small red light. Improvements such as these make a great difference to the ease and pleasure of observing out in the field. It is also essential to have a white light, either a torch (flashlight) or one run from the car battery on a long cable, to help find those vital accessories that always seem to get dropped. However, if you are going to be observing with other astronomers around, be careful when using it or you could make yourself very unpopular.

A good magnet is well worth having, to help retrieve screws and bolts that you drop when fumbling with cold hands to fix a telescope to its mount. It is worth painting the heads of screws and bolts red or yellow. A black or even a shiny object has a knack of disappearing in the grass when it falls to the ground.

I shall gloss over some of the other paraphernalia of observing in the country, such as warm clothing, insect repellent, refreshments, and so on. There is also the matter of accurate polar alignment, which is particularly important if you want to attempt photography. Many observing guides tell you how to do this; all I shall add is that if you have a mounting with a polar alignment system, take along a piece of old carpet which you can lie on. This will also come in useful when you are trying to line up your camera on a particular area for piggy-back photography.

It sounds obvious to say "make a list of everything you will need," and I am the first to ignore this advice. But if you have no list, there is one item you should always double-check if you are using a German mounting – the counterbalance. These are probably the most easily forgotten items, followed by the removable shaft on which the counter-balance slides.

Lastly, do not forget that if you are hoping for good seeing, there is no point in observing in a line of sight that passes over the car, for its cooling engine will continue to give off heat long after it has been turned off. More obviously, avoid a line of sight that passes near the exhaust if the engine is running. And if you leave your camera on a driven platform behind the car, carrying out a time exposure while you sit inside with the engine running to keep yourself warm, you may well find that the camera is immersed in a cloud of condensation from the exhaust.

The portable telescope

Whether you travel to your dark site on foot, by car or by plane, you may well be taking a telescope with you. Any telescope which is not permanently fixed on a massive mounting is a compromise in one way or another, and you have to decide for yourself what suits your own circumstances. There are various possibilities.

To my mind, a telescope is portable only if you can easily pick it up, mounting and all, with one hand. Everything else is merely trans-portable. Horace Dall was the expert at truly portable telescopes. His finest was a 150 mm (6-inch) reflector which he could conveniently carry in a raincoat pocket. This is no exaggeration. He traveled all over the world with it, and it was never noticed by customs officials, though that was in the days before metal detectors at airports. The

◀ *Fig. 7.8 Telescope wizard Horace Dall (1901–86) with his ultra-portable 150 mm Dall-Kirkham design reflector, which folded down to a package little larger than the mirror – the ultimate in traveling telescopes.*

design was a "thin-mirror Dall-Kirkham." The secondary was carried on an arm which moved along a short square-section beam for focusing. It fitted easily to a small photographic tripod (Figure 7.8). To make such a device commercially would be prohibitively expensive, and most of us have to settle for more mundane instruments.

Traveling by air is by far the quickest way, and often the cheapest way, of getting to dark skies. On many international flights there is a baggage limit of 20 kg (44 lb), which does not go very far where equipment is concerned. A basic 200 mm (8-inch) SCT will take you to the limit by itself. Furthermore, the hassle of air travel is increased the more you carry: you have more to look after, and there is always the risk that an item will get lost, stolen, or sent to the wrong place and not returned to you until it is too late. So, many people would prefer to take their telescope with them as cabin baggage.

If you are over the baggage limit you may be charged a considerable excess, which may be based on the first-class fare rather than your economy-class ticket. The best way is to travel with other passengers who are carrying less than the limit, and make sure that you check in at the same time as one another, because airlines (other than the penny-pinching budget ones) generally allow groups traveling together to pool their weight allowances.

For years the Meade ETX range has been the leading choice for portable instruments, particularly the ETX-90 Maksutov, which weighs in at 3.5 kg (7.7 lb) plus an optional tripod at 5 kg (11 lb), and is small enough to fit in standard cabin baggage. Today there are more options, such as 127 mm or larger catadioptric telescopes from Sky-Watcher or Celestron, and cheaper 90 mm scopes or alternative mounts (Figure 7.9). There are general-purpose Go To mounts, on which you can mount a variety of telescopes, such as the Sky-Watcher or iOptron cube mounts. Shipping a larger instrument is possible: planetary imagers Damian Peach and Dave Tyler each took 14-inch

Celestrons from the UK to Barbados for an imaging trip! The baggage weight was shared between their families who went with them, avoiding excess costs. At the other end of the scale, if you just want to take some photos of deep-sky wonders then a portable driven mount such as the AstroTrac or Vixen Polarie is all you need. Your DSLR or CSC camera with a telephoto lens on such a mount will give you beautiful wide-field views of nebulae and star clusters, while a wide-angle or even a fisheye lens will take brilliant views of the Milky Way.

Catadioptric instruments such as Maksutovs and Go To mounts are fine for general viewing if you need to carry the instrument by plane. Don't try to save weight by aiming to use a camera tripod with its pan and tilt head for carrying these telescopes – they have long focal lengths and the movements may not be smooth enough.

If you are not flying and just want get to dark skies in the car then aperture is the answer. Several manufacturers now make large Dobsonians with framework or truss tubes which dismantle. At star parties you will see 450 mm (18-inch) or larger instruments made in this way being brought in ordinary cars. While even a 250 mm (10-inch) Newtonian telescope may not fit easily in your car, occupying the whole back seat, a similar-sized truss-tube Dob could leave room for the passengers and their luggage as well.

But most large instruments are a lot heavier and a good deal more cumbersome, and most people would not want to carry them around regularly. The most useful telescope is the one you can observe with easily, as William Herschel found when he built his giant 48-inch (1.2-meter) reflector, then the largest in the world. It was so unwieldy in use that he preferred his smaller instruments most of the time.

▶ *Fig. 7.9 The Sky-Watcher Skymax-127 Synscan, or its Celestron equivalent, such as the NexStar 5, gives a reasonably large aperture in a compact Go To instrument suitable for traveling. The weight with tripod is 8 kg (18 lb).*

8 · INDOOR ASTRONOMY

With the best will in the world, the amateur astronomer cannot observe every night. Indeed, for the city-bound observer good opportunities to see the sky are infrequent, leaving plenty of time on cloudy nights, or when the conditions are otherwise unsuitable, for pursuing astronomy from the warmth and comfort of indoors. And there are those whose enthusiasm does not extend to observing sessions on every clear night.

Through the glass darkly

If your reason for not going out is that it is just too cold, well, you can actually do some observing from indoors. And if you are incapacitated and unable to go outside, observing through a window may be your

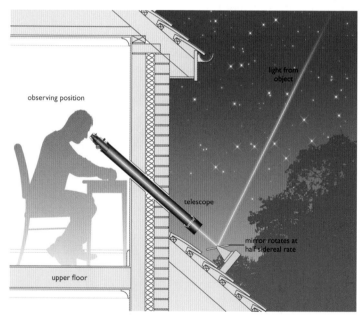

▲Fig. 8.1 For indoor comfort while observing, arrange a silvered flat mirror below a south-facing window so that it reflects light into a telescope set such that its eyepiece is at a convenient viewing angle. The drawbacks of this system are that only a comparatively small part of the sky can be observed, and a suitable flat mirror (which should be rather larger than the telescope's objective, to allow for observing at an extreme angle of reflection) of sufficiently high optical quality is likely to be expensive.

only option. Every textbook on amateur astronomy tells you that it is a big mistake to observe through an open window, because the turbulence created as air escapes from your warm room into the cold night will destroy the seeing. Of course, they are right – but only up to a point. They don't usually mention observing through a *closed* window.

First, this question of seeing. As I pointed out in Chapter 2, seeing is all-important to the planetary observer. If you want good seeing, you must avoid temperature variations in your vicinity. Even if you open the window well before you start to observe, the room will probably still retain a large amount of heat, even in winter. But I daresay most observers have pointed a telescope out of the window at some stage, and found that the seeing is not that bad. There is something to be said for it – a warm and comfortable observer can make much better observations.

In fact, as long as you insulate the interior from the exterior sufficiently there is little wrong with indoor observing. Astronomy author James Muirden once made a coelostat so that he could observe the planets from his south-facing study window. A coelostat is simply a flat mirror that reflects the light from the object under study into a fixed telescope. The mirror is rotated slowly in such a way as to compensate for the apparent rotation of the celestial sphere. In the house where James was living at the time, the pitched roof of a porch was just below his study window, ideally positioned for the coelostat (see Figure 8.1). A small refractor was angled downward toward the mirror. The eyepiece end was inside the study, and one looked into it as if it were a microscope eyepiece. I have never before or since observed in such comfort.

The main drawbacks with this system are that the light is dimmed somewhat by the mirror, and the image is laterally reversed. Finding planets was fairly easy, but it would be more difficult to locate deep-sky objects. Also, observations are limited to a smaller part of the sky than with a conventional telescope, depending on where the coelostat is sited. The main expense is the flat mirror – it has to be of high optical quality, which can make it almost as expensive as a similar-sized paraboloidal mirror for a telescope.

If that all seems too much bother, you might care to try observing through the glass of a closed window, particularly if you are using binoculars. Modern float glass can be of quite good optical quality, and while it is not good enough for use as a flat mirror, it can have comparatively small effects on transmitted light. As long as the power is not too high and you are not observing at too shallow an angle to the glass, the image can be surprisingly good. A little contrast and light are lost, but this matters only if you are attempting very difficult

◀ *Fig. 8.2 The late George Alcock's record of novae and comet discoveries is all the more impressive because of his unpromising site near Peterborough, UK. For many years there were high levels of industrial pollution from nearby brickworks, now disused. When into his eighties, George preferred to carry out his binocular sweeps through double-glazing.*

observations. With low powers you will notice almost no difference when observing through a good window.

Though your view is restricted, it is still possible to carry out useful work through a window. Variable-star observing using binoculars is one possibility. There are even examples of a discovery having been made through a window. In 1961, comet expert Michael Candy was testing a telescope by pointing it through a bedroom window when he spotted a fuzzy object that should not have been in the field of view. It turned out to be a new comet, now designated Comet Candy 1961 II. Then in 1991 the late nova- and comet-hunter George Alcock (Figure 8.2) discovered his sixth nova, in Hercules. He used 10 × 50 binoculars through a double-glazed window to spot the fifth-magnitude newcomer! This was hardly a chance discovery – in the previous 27 months he had spent a total of 553 hours searching for just such an object. He spotted the nova at 4.35 a.m., the sky having cleared just half an hour previously. He also discovered Comet IRAS–Araki–Alcock while observing through double-glazing. But George is no longer with us, and I doubt that even he would be a match for today's automated nova and comet searches.

Observing in other ways

Radio astronomy is a major part of the professional scene, but it has had little impact on amateurs. It mostly lacks the romantic appeal of visual astronomy – there are few dramatic images, just records of radio intensity. You might also be misled by thinking about the giant dishes used by professionals to pick up the feeble signals from the farthest reaches of the Universe. But there are radio astronomy projects that use quite simple equipment, often home-made or at least home-assembled, which can be as much fun as any other type of amateur observing if it's the sort of thing that appeals to you.

Meteors can be observed by radio with the simplest of equipment – an ordinary FM radio, preferably one with digital tuning. It also helps to have an additional outdoor aerial (antenna), ideally a directional Yagi-type array. The technique is simply to choose a radio station which transmits on a frequency well separated from those that you can receive locally, yet is too distant for you to receive directly – maybe one 1,300–2,000 km (800–1,300 miles) away. The frequency must otherwise be clear. A digital tuner helps you to tune to the correct frequency without being able to hear the station. Normally, you will not receive signals from the station, but if a meteor dashes through a region of the upper atmosphere between you and the radio transmitter, the signal will be reflected by the meteor's ionization train, and you will suddenly pick up a short, coherent burst from the station. The burst, known as a "ping," may last only for the briefest instant, but if a meteor leaves a substantial train you could pick up the distant station for several seconds.

It helps to know of a variety of stations in different directions, because the best reflections occur when the meteor is broadside on to both you and the station. This means finding a station at right angles to the direction of the radiant. On a night when a rich meteor shower is near maximum you may detect up to 100 meteors an hour in this way. This technique is useful on those frustrating occasions when a very high hourly rate is predicted, but it is either daylight or cloudy. Then you can keep tabs on the shower's activity from your armchair. (Further details are given at www.skyscan.ca/meteor_radio_detection.htm).

The ready availability of electronic devices, satellite dishes and receivers has widened the range of projects that you could undertake. The British Astronomical Association's Radio Astronomy Group lists several on its website (www.britastro.org/radio), from picking up radio storms on Jupiter to detecting hydrogen emission from the Milky Way.

This sort of thing is fairly well removed from standard amateur astronomy activities. You will not usually find the apparatus advertised in the astronomy magazines, so unless you are already familiar with electronics and computers you will need to put in some effort. Radio observations, however, are likely to be hindered by nearby sources of electrical interference. Your neighbors' electrical appliances, badly suppressed car ignitions, electric trains, and maybe even central-heating controllers could all disturb your records. These disturbances can happen in country areas too, but they will be worse in more densely populated districts. Paul Hyde, of the BAA's Radio Astronomy

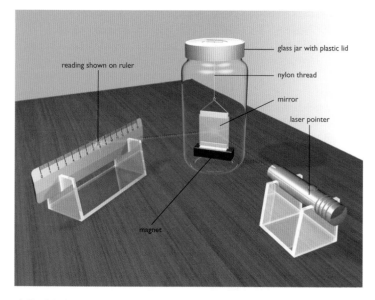

▲ *Fig. 8.3 A jamjar magnetometer is simple to set up and use. Small variations in the Earth's magnetic field are made* *visible by the "leverage" provided by the reflection of a light beam, in this case a laser pointer.*

Group, reports that, "The latest issue is with a minority of solar panel installations that broadcast high levels of broadband interference. My neighbor has a solar panel system which is not a problem but two doors further down another system has rendered my location unusable during sunny conditions."

While it is possible to observe other bodies in space at radio wavelengths, the level of signal from anything except the Milky Way and powerful sources such as the Crab Nebula is too small to be detected without a rather large collecting dish.

There is, however, a very simple means of monitoring a different form of activity, and the basic apparatus is almost laughable – part of it is a jamjar (Figure 8.3). A bar magnet suspended so that it can swing freely will align itself with the Earth's magnetic field. The streams of solar particles that cause aurorae also produce changes in the Earth's magnetic field which are detectable at geomagnetic latitudes lower than those from which the aurorae are visible. These changes show up as tiny shifts in the magnet's position, which would be hard to detect just by looking at the magnet, but there is a simple method of making them easy to see. Attach a small household mirror to the magnet, shine a light, or these days a laser pointer, on to it, and watch the reflection of the light beam on a screen some distance away. (Further details are given at www.britastro.org/aurora/jamjar.htm.) During a magnetic storm the beam will shift visibly by several millimeters in a matter of minutes.

Joining forces

Astronomers, like galaxies, frequently gather in local groups. If you live in a big enough town, you may find that there is a club or society that meets near you, so you probably have the advantage over someone who enjoys pitch-black skies yet lives miles from any other astronomers. At a local society you will discover the wide range of amateur types – the experienced observers, the eclipse chasers, the armchair astronomers, the telescope nuts, and the out-and-out loonies. But most will be ordinary folk like yourself, maybe without much scientific background, and simply keen to find out more about astronomy. Some will be more advanced than others, but do not worry that your own knowledge is inadequate. A good society caters for many levels and tastes. Give yourself a few meetings to fit in, for your first visit might coincide with the most technical or the worst talk of the year.

I have seen complaints that such-and-such a society seems unfriendly, and despite attending several meetings no one spoke to the new member. It seems to me that you do have to make a bit of an effort. The other side of the coin is that some newcomers may prefer not to be put on the spot by a conversation in which the gaps in their knowledge are exposed. So if you want help, just ask. Most societies will have members who can help you with buying equipment, or advise on astrophotography, for example. You may even find that you have some particular knowledge or experience that enables you to help others. I have been a member of many societies, and have helped run more than a few. They can be fun and rewarding, and I would encourage anyone who wants to get more out of the subject to join one. But some would rather just get on with astronomy. Many people are by nature not joiners, in which case, fine.

There are larger organizations as well. Most countries have national amateur societies, such as the British Astronomical Association and the Society for Popular Astronomy in Britain, the Royal Astronomical Society of Canada, and the Royal Astronomical Society of New Zealand. In the US there is no single organization, though the Astronomical League coordinates the activities of local societies, a task performed in Britain by the Federation of Astronomical Societies. The US has specialist organizations such as the Association of Lunar and Planetary Observers (ALPO) and the American Association of

Variable Star Observers (AAVSO), both of which have members around the world, including professionals.

There are also societies which, though based in one place, have an international membership. The Webb Deep-Sky Society, devoted to deep-sky observing, is one of these. Another is The Astronomer, also based in the UK, actually the name of a monthly magazine sent out by post or email, though its subscribers make up an informal network of observers who often provide an invaluable service in checking new discoveries before they are officially announced. The magazine simply lists observations made the previous month, with occasional articles on observing techniques, while a circular service available by e-mail gives timely warning of a wide range of discoveries. The International Meteor Organization, based in Belgium, also has a worldwide membership. Societies such as these do have meetings, but their value is that they provide a means of communication rather than acting as a social club.

Many local societies have their own observatories, some of which are equipped with meeting rooms, libraries, and even overnight accommodation. Rather than spend a large sum of money equipping yourself with a large telescope and all its ancillary equipment, you can join the society and use its telescope instead. But I would always advise having your own telescope, even if it is just a basic 150 mm (6-inch) reflector. The best telescope is the one that gets used the most.

Astronomy on the web

Today, if we want to know anything, we turn to the web. There are thousands of astronomical websites available, and it's very difficult to pick out just a few of the most important or useful ones other than those that I've mentioned in the text already. If I were to give even a limited list here, it would probably start to get out of date within a year or two. In most cases, a search engine will find what you want anyway.

But the web is more than just websites. These days, forums are a wonderful resource for getting practical information. People share their user experiences of observing, telescopes, cameras, filters... you name it, someone, somewhere, has commented on it. Some forums are huge, with thousands of members – though I know only too well that a large number of such registrations are actually bogus, designed solely to gain a web presence via the member list. Even so, you will quickly get all shades of experience and advice, some of it invaluable, and some of it rubbish. Some sites are carefully moderated, so the more blatant rudeness, wrong advice or just plain advertising is filtered out, but by and large you do need to be wary about what people say. The anonymous nature of forums can be a problem, as people know that they can't be traced, and you'll soon notice that some comments are rather more, well, forthright than you would normally expect in a conversation. You just have to take all things into account – how many other users seem to know the member whose views you are reading, how often have they posted, where do they live, and do some users know their real name and maybe meet them in person? I always tend to disregard the views of somone who feels the need to rubbish the things that others say.

Citizen science

For the urban astronomer who wants to take their hobby further, yet finds that the light-polluted skies are too offputting, there are now ways in which you can get involved online. This started with the SETI@home screensaver (setiathome.berkeley.edu), in which people in effect donate their computer power to the search for extra-terrestrial life (SETI). You download a package of data, then let your computer analyze it while you aren't using it (Figure 8.4). The whole process takes place automatically, and you know that you are helping with the search for radio signals from other civilizations.

Particularly successful has been the Galaxy Zoo project (www. galaxyzoo.org), which was set up to help analyze the vast amount of

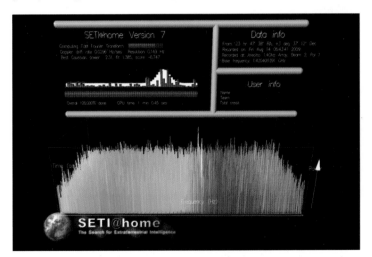

▲ *Fig. 8.4 This SETI@home screensaver kicks in whenever you are not using the computer and automatically downloads* *and processes data from radio telescopes searching for anomalous signals that could be from alien civilizations.*

◀▲ *Fig. 8.5 Dutch schoolteacher Hanny van Arkel hit the limelight when she spotted a strange green object (below the galaxy at right) on images she was checking for Galaxy Zoo. It turned out to be a new class of object, and is now known as Hanny's Voorwerp (Hanny's Object). Here she chats with Dr Brian May (left) and Professor Brian Cox.*

data coming from deep-sky surveys. Participants are asked to study and classify galaxy images sent to them, according to guidelines. In this case, you get to see the actual data rather than allowing your computer to do the work, and your input can lead to actual discoveries. One exciting feature is that because the individual galaxies have been selected automatically by a computer, you could be the first person actually to see and classify them (Figure 8.5). You might think that the classification could also be done by computer, but actually the human brain is far better

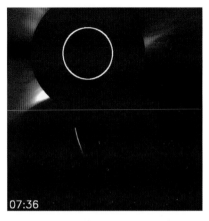

▶ *Fig. 8.6 Another comet plunges headlong into the Sun, seen only on images from the SOHO spacecraft. As of 2014, over 2,500 comets had been discovered on SOHO images.*

at seeing shapes and patterns than a computer – try training a computer to distinguish a cat from a dog, for example. The project is making real discoveries about the distribution of galaxies in the Universe, just on the basis of these straightforward classifications. Some new types of galaxy have been uncovered, including "green-pea galaxies" and red spirals, and numerous research papers have been published as a result of the work done by ordinary people sitting at their computers. Springing out of the Galaxy Zoo project is Zooniverse (www.zooniverse.org), which offers more projects that you can participate in.

There is masses of data now available on the web, and sometimes the opportunities that it offers are missed by the people who gathered it. The SOHO comets are a good example. The solar images taken from the SOHO spacecraft since it began its mission in 1996 were found to contain faint comets, previously unseen, that became visible only as they plunged close to the Sun, often evaporating completely as they did so. But these were not the primary purpose of SOHO, and they were largely ignored until, in 2000, an amateur astronomer named Michael Oates attended a meeting of the Society for Popular Astronomy in London and heard that it was possible to discover comets in this way. Within days he had found one for himself in the SOHO data, and went on to find many more in the archives. I was at the same meeting, but Mike was the one who got on with the job! Today keen-eyed Chinese are top of the list of SOHO comet discoverers, though the discoveries are all attributed to SOHO rather than the individual (Figure 8.6).

These initiatives come under the general classification of "citizen science," and a web search on this topic could lead you to find new projects.

DEEP-SKY OBJECTS VISIBLE FROM URBAN AREAS

Unlike most other lists of deep-sky objects, these have been compiled especially for the urban observer, with the help of David Frydman, Stanley Gleeson, Graham Long, Ian Ridpath, Steve Smith and Richard Westwood. Positions given are for epoch 2000.0. All these objects have been seen from urban areas, but not necessarily from city centers. Some may require ideal conditions and a fairly large aperture, and others may be beyond your reach. On the other hand, you may be able to see objects not listed here. Some general visibility guidelines are:

- Any open cluster brighter than overall magnitude 7 is probably well seen if it is above 20° elevation.
- Any globular cluster brighter than magnitude 8 is well seen if above 45° elevation.
- Planetary nebulae can be seen if they are small and above 45° elevation, particularly if you use a filter. Some need a high magnification, which makes them hard to locate.
- Galaxies are usually difficult from cities unless they are higher than 65°, especially if they are seen face-on rather than edge-on. Filters do not usually help.
- Few diffuse nebulae are well seen from towns, though a filter can improve their visibility.
- Double stars are well seen above 20°.
- The larger your telescope, the fainter the objects you can see – even from bright locations.

The top ten city objects
No definitive choice is possible, because even bright objects can be invisible when low in the city sky. Observers in Britain and Canada get a poor view of M6 and M7, for example, while an observer farther south would rank M13 a very poor second to Omega Centauri. So I have divided the sky into three bands: high northern latitudes, including much of northern Europe and Canada; mid-northern latitudes, including the southern US and farther south; and the southern hemisphere. I have included some objects just because people want to see them, rather than because they are particularly spectacular. All these objects should be observable from city skies without the use of a filter. I have put them roughly in order of interest, although the choice is a little subjective and others will have their own ideas.

1. TOP TEN SHOWPIECES VISIBLE FROM HIGH NORTHERN LATITUDES

Designation	RA	Dec	Object	Constellation	Comments
M45	03h 47m	+24° 07'	Loose cluster	Taurus	Pleiades. Excellent in binoculars.
M42	05h 35m	–05° 30'	Diffuse nebula	Orion	Orion Nebula. Visible with the naked eye even from the city; gets better with more aperture.
NGC 869 and 884	02h 20m	+57° 08'	Open clusters	Perseus	Double Cluster/h and Chi. Naked-eye object when high up; low power or binoculars.
M44	08h 40m	+20° 00'	Open cluster	Cancer	Praesepe or Beehive. Good in binoculars.
M27	20h 00m	+22° 43'	Planetary nebula	Vulpecula	Dumbbell Nebula. Can just be seen with binoculars.
M57	18h 54m	+33° 02'	Planetary nebula	Lyra	Ring Nebula. Needs a power of 50 or more, but a favorite object for small telescopes.
Beta Cygni	19h 31m	+27° 58'	Double star	Cygnus	Albireo. Contrasting colors make this one of the sky's best doubles.
M31	00h 43m	+41° 16'	Galaxy	Andromeda	Andromeda Galaxy. The brightest galaxy; visible to the naked eye and worth a look with any optical aid you have.
M13	16h 42m	+36° 28'	Globular cluster	Hercules	Barely visible in binoculars. Use a moderate power on a telescope to resolve stars.

2. TOP TEN SHOWPIECES VISIBLE FROM MID-NORTHERN LATITUDES

Designation	RA	Dec	Object	Constellation	Comments
M6	17h 40m	–32° 13'	Open cluster	Scorpius	Hard to see from Britain and Canada.
M7	17h 54m	–34° 49'	Open cluster	Scorpius	Hard to see from Britain and Canada.
M8	18h 04m	–24° 23'	Diffuse nebula	Sagittarius	Lagoon Nebula. Visible in binoculars.
M11	18h 51m	–06° 16'	Open cluster	Scutum	Wild Duck. Visible in binoculars, but needs more aperture for best views.
NGC 253	00h 48m	–25° 17'	Galaxy	Sculptor	Edge-on spiral. Hard to see from Britain and Canada.
Omega Centauri	13h 27m	–47° 29'	Globular cluster	Centaurus	Good in binoculars; gets better with increasing aperture. Best seen from the southern hemisphere.
M17	18h 21m	–16° 11'	Diffuse nebula	Sagittarius	Omega Nebula.
M41	06h 47m	–20° 44'	Open cluster	Canis Major	Good in binoculars.
NGC 6231	16h 54m	–41° 48'	Open cluster	Scorpius	Best seen from the southern hemisphere.
NGC 2477	07h 52m	–38° 33'	Open cluster	Puppis	Best seen from the southern hemisphere.

3. TOP TEN SOUTHERN SHOWPIECES

Designation	RA	Dec	Object	Constellation	Comments
Large Magellanic Cloud (LMC)	05h 30m	–69°	Galaxy	Dorado	Naked-eye and binocular object.
Small Magellanic Cloud (SMC)	01h 00m	–72°	Galaxy	Tucana	Naked-eye and binocular object.
47 Tucanae	00h 24m	–72° 05'	Globular cluster	Tucana	Very close to SMC.
NGC 3372	10h 45m	–59° 50'	Diffuse nebula	Carina	Eta Carinae Nebula. Naked-eye object.
NGC 2070	05h 39m	–69° 06'	Diffuse nebula	Dorado	Tarantula Nebula. In LMC. Bright and complex, enhanced by filter.
NGC 3532	11h 06m	–58° 40'	Open cluster	Carina	–
IC 2602	10h 43m	–64° 24'	Open cluster	Carina	Southern Pleiades. Needs low power.
NGC 4755	12h 54m	–60° 20'	Open cluster	Crux	Jewel Box. Disappointing in binoculars. Needs some aperture and power to show a brilliant range of star colors.
NGC 2516	07h 58m	–60° 52'	Open cluster	Carina	–
NGC 3132	10h 38m	–40° 26'	Planetary nebula	Vela	Brighter than the Ring Nebula; needs similar power. Prominent central star.

4. TOP TEN OBJECTS FOR IMAGERS

These are all comparatively easy to image using a filter and telephoto lens, yet virtually impossible to see from the city without a filter. In addition, many other diffuse nebulae mentioned in the lists on these pages can be photographed.

Designation	RA	Dec	Object	Constellation	Comments
NGC 7000	21h 00m	+44° 20'	Diffuse nebula	Cygnus	North America Nebula. Popular object, but needs good red sensitivity and contrast.
NGC 2237	06h 30m	+05° 03'	Diffuse nebula	Monoceros	Rosette Nebula.
NGC 1499	04h 01m	+36° 37'	Diffuse nebula	Perseus	California Nebula. Large and faint.
B33, IC 434	05h 41m	−02° 24'	Dark nebula	Orion	Horsehead Nebula. Needs a long focal length to reveal the actual Horsehead, B33, though the bright nebula IC 434 can be photographed with a standard lens.
NGC 2024	05h 42m	−01° 51'	Diffuse nebula	Orion	Flame Nebula. Just east of Zeta Orionis.
NGC 6960/ 6992	20h 56m	+31° 43'	Supernova remnant	Cygnus	Veil Nebula. The two nebulae form arcs of a circle. Position given here is for 6992, the brightest part. Visible on good nights with filter.
NGC 2264	06h 41m	+09° 53'	Cluster, dark nebula	Monoceros	Cone Nebula. As with the Horsehead, needs long focal length to show the dark nebula.
–	05h 50m	00° 00'	Supernova remnant	Orion	Barnard's Loop. Faint, but shows up because of its size (about 10°).
NGC 281	00h 53m	+56° 37'	Diffuse nebula	Cassiopeia	Many other nebulous objects nearby.
–	20h 22m	+40° 15'	Diffuse nebulae	Cygnus	Gamma Cygni. Whole area wreathed with nebulosity.

5. TOP TEN DOUBLE STARS

Designation	RA	Dec	Object	Comments
Beta Cygni	19h 31m	+27° 58'	3.1, 5.1	Albireo. Separation 34.6 arc seconds.
Gamma Andromedae	02h 04m	+42° 20'	2.3, 5.1	Almach. Separation 9.6 arc seconds.
Epsilon Lyrae	18h 44m	+39° 40'	4.7, 4.6	The Double Double. Main star separation 210.5 arc seconds; its components are 2.3 and 2.4 arc seconds.
Theta Orionis	05h 35m	−05° 23'	5.1, 6.7	Trapezium; the two other stars are variable.
Epsilon Boötis	14h 45m	+27° 04'	2.7, 5.1	Pulcherrima. Separation 2.8 arc seconds.
Alpha Crucis	12h 27m	−63° 06'	1.3, 1.7	Acrux. Separation 4.4 arc seconds.
Alpha Centauri	14h 40m	−60° 50'	0.0, 1.3	Rigil Kentaurus. Period 80 years, so variable separation.
Sigma Orionis	05h 39m	−02° 36'	4.0, 6.0	Multiple star.
Eta Cassiopeiae	00h 49m	+57° 49'	3.5, 7.5	Separation 13 arc seconds.
Delta Cephei	22h 29m	+58° 25'	3.6, 6.3	The original Cepheid star. The primary varies from magnitude 3.6 to 4.3.

6. NOW TRY THESE

Having whetted your appetite on the sky's showpieces, you will want to look further. All the objects in this list are visible from city locations; there are many others besides, particularly in the southern sky. Many of these objects can be picked out with binoculars, but you need a telescope to do them proper justice – and the larger it is, the better the view.

Designation	RA	Dec	Object	Constellation	Comments
NGC 55	00h 15m	−39° 11'	Galaxy	Sculptor	–
M110	00h 40m	+41° 41'	Galaxy	Andromeda	See comments for M32. Also known as NGC 205.
M32	00h 43m	+40° 52'	Galaxy	Andromeda	Famous mostly as a companion to M31, otherwise unimpressive.
NGC 404	01h 09m	+35° 43'	Galaxy	Andromeda	Close to Beta Andromedae, which must be kept out of the field of view. A test of the contrast of your optics.
NGC 457	01h 19m	+58° 20'	Open cluster	Cassiopeia	Owl Cluster. Fine contrast of color, with a fairly compact center.
M103	01h 33m	+60° 42'	Open cluster	Cassiopeia	–
M33	01h 34m	+30° 39'	Galaxy	Triangulum	Pinwheel Galaxy. Very difficult in light-polluted skies.

Designation	RA	Dec	Object	Constellation	Comments
M76	01h 42m	+51° 34'	Planetary nebula	Perseus	Little Dumbbell. Needs a high power.
NGC 752	01h 58m	+37° 41'	Loose cluster	Andromeda	Large but unimpressive with binoculars; low power on larger telescope needed.
M34	02h 42m	+42° 47'	Open cluster	Perseus	Includes several pairs and triple stars.
Melotte 20	03h 20m	+50°	Loose cluster	Perseus	Alpha Persei cluster.
NGC 1502	04h 08m	+62° 20'	Open cluster	Camelopardalis	Within Kemble's Cascade.
NGC 1528	04h 15m	+51° 14'	Open cluster	Perseus	–
–	04h 30m	+17°	Loose cluster	Taurus	Hyades. Too large (about 5°) to be seen in any instrument but binoculars.
NGC 1647	04h 46m	+19° 04'	Open cluster	Taurus	Needs low power on a large aperture for the best views. Near the Hyades.
NGC 1758	05h 04m	+23° 46'	Open cluster	Taurus	Very large cluster.
M38	05h 29m	+35° 50'	Open cluster	Auriga	Greater mix of bright and faint stars than M34 or M36. Look for the fainter NGC 1907 near its edge.
M1	05h 35m	+22° 01'	Supernova remnant	Taurus	Crab Nebula. Needs moderate power. Not spectacular visually without a filter.
NGC 1981	05h 35m	–04° 26'	Open cluster	Orion	–
M36	05h 36m	+34° 08'	Open cluster	Auriga	Together with M37 and M38, makes a fine sight in binoculars.
NGC 1977	05h 36m	–04° 52'	Diffuse nebula	Orion	–
M43	05h 36m	–05° 16'	Diffuse nebula	Orion	Northeast part of Orion Nebula.
NGC 2017	05h 39m	–17° 51'	Multiple star	Lepus	–
M78	05h 47m	+00° 03'	Diffuse nebula	Orion	Comet-like.
M37	05h 52m	+32° 33'	Open cluster	Auriga	Needs darker sky than M36 and M38.
M35	06h 09m	+24° 20'	Open cluster	Gemini	Compact cluster NGC 2158 is nearby, visible in larger telescopes.
NGC 2244	06h 32m	+04° 52'	Open cluster	Monoceros	Surrounded by the Rosette Nebula, NGC 2237 (see photographic list).
NGC 2251	06h 35m	+08° 22'	Open cluster	Monoceros	Linear star patterns.
NGC 2264	06h 41m	+09° 53'	Open cluster	Monoceros	–
M50	07h 03m	–08° 20'	Open cluster	Monoceros	–
NGC 2362	07h 19m	–24° 57'	Open cluster	Canis Major	Needs moderate power.
NGC 2392	07h 29m	+20° 55'	Planetary nebula	Gemini	Eskimo Nebula, so-called because of apparent features within the nebula.
M47	07h 37m	–14° 30'	Open cluster	Puppis	–
M46	07h 42m	–14° 49'	Open cluster	Puppis	–
NGC 2451	07h 45m	–37° 58'	Open cluster	Puppis	–
M67	08h 50m	+11° 49'	Open cluster	Cancer	–
NGC 2903	09h 32m	+21° 30'	Galaxy	Leo	One that Messier missed.
M81	09h 56m	+69° 04'	Galaxy	Ursa Major	A binocular object in good skies. In the same field as M82.
M82	09h 56m	+69° 41'	Galaxy	Ursa Major	Better from city skies than nearby M81.
NGC 3242	10h 25m	–18° 38'	Planetary nebula	Hydra	The Ghost of Jupiter, so-called because of its size and elliptical shape.
M65	11h 19m	+13° 05'	Galaxy	Leo	Noticeably elliptical.
M66	11h 20m	+12° 59'	Galaxy	Leo	Less tilted than M65. Nearby stars make M65 and M66 easy to find.
NGC 3918	11h 50m	–57° 11'	Planetary nebula	Centaurus	A small but bright planetary, blue in color.
Melotte 111	12h 25m	+26°	Loose cluster	Coma Berenices	Coma Star Cluster. Like the Hyades, too large for a telescope.
M104	12h 40m	–11° 37'	Galaxy	Virgo	–

Designation	RA	Dec	Object	Constellation	Comments
NGC 4945	13h 05m	−49° 28'	Galaxy	Centaurus	Very faint from city site; a large, long object.
M53	13h 13m	+18° 10'	Globular cluster	Coma Berenices	–
NGC 5128	13h 26m	−43° 01'	Galaxy	Centaurus	Centaurus A. A large, slightly oblong blob with just a hint of a dust lane.
M3	13h 42m	+28° 23'	Globular cluster	Canes Venatici	–
NGC 5307	13h 51m	−51° 12'	Planetary nebula	Centaurus	Difficult, very small and not much larger than a star. Filter needed.
M101	14h 03m	+54° 21'	Galaxy	Ursa Major	–
IC 4406	14h 22m	−44° 09'	Planetary nebula	Lupus	Small object, relatively bright and circular.
M5	15h 19m	+02° 05'	Globular cluster	Serpens	–
NGC 5986	15h 46m	−37° 47'	Globular cluster	Lupus	Small object, fairly faint. One star at its edge is relatively bright.
M4	16h 24m	−26° 32'	Globular cluster	Scorpius	–
NGC 6210	16h 45m	+23° 49'	Planetary nebula	Hercules	–
M12	16h 47m	−01° 57'	Globular cluster	Ophiuchus	–
M10	16h 57m	−04° 06'	Globular cluster	Ophiuchus	–
M92	17h 17m	+43° 08'	Globular cluster	Hercules	–
M14	17h 38m	−03° 15'	Globular cluster	Ophiuchus	–
NGC 6397	17h 41m	−53° 40'	Globular cluster	Ara	Easy object, a fairly open cluster but with condensed nucleus.
IC 4665	17h 46m	+05° 43'	Loose cluster	Ophiuchus	Best seen in binoculars – a loose scattering of stars.
NGC 6543	17h 59m	+66° 38'	Planetary nebula	Draco	–
M16	18h 19m	−13° 47'	Cluster and nebula	Serpens	Eagle Nebula.
NGC 6633	18h 28m	+06° 34'	Open cluster	Ophiuchus	Needs a low power; bright, yellowish stars.
M22	18h 36m	−23° 54'	Globular cluster	Sagittarius	–
M56	19h 17m	+30° 11'	Globular cluster	Lyra	–
Collinder 399	19h 25m	+20° 11'	Loose cluster	Vulpecula	Coathanger. Also known as Brocchi's Cluster.
NGC 6826	19h 45m	+50° 31'	Planetary nebula	Cygnus	The Blinking Planetary, so-called because it seems to blink on and off as the eye is attracted to the bright central star.
M71	19h 54m	+18° 47'	Globular cluster	Sagittarius	–
M29	20h 24m	+38° 32'	Open cluster	Cygnus	Needs low power.
IC 5067	20h 48m	+44° 22'	Diffuse nebula	Cygnus	Pelican Nebula. Needs filter.
NGC 7006	21h 01m	+16° 11'	Globular cluster	Delphinus	–
NGC 7009	21h 04m	−11° 22'	Planetary nebula	Aquarius	Saturn Nebula.
NGC 7027	21h 07m	+42° 14'	Planetary nebula	Cygnus	–
M15	21h 30m	+12° 10'	Globular cluster	Pegasus	–
M39	21h 32m	+48° 26'	Open cluster	Cygnus	–
M2	21h 34m	−00° 49'	Globular cluster	Aquarius	–
NGC 5150	21h 59m	−39° 25'	Planetary nebula	Grus	Very faint, reasonable size. Filter helps.
NGC 7293	22h 30m	−20° 48'	Planetary nebula	Aquarius	Helix Nebula. Needs filter.
NGC 7331	22h 37m	+34° 25'	Galaxy	Pegasus	–
NGC 7510	23h 12m	+60° 34'	Open cluster	Cepheus	Strange bar shape.
M52	23h 24m	+61° 35'	Open cluster	Cassiopeia	Similar to M37, but maybe easier.
NGC 7662	23h 26m	+42° 33'	Planetary nebula	Andromeda	Needs high power.
NGC 7789	23h 57m	+56° 44'	Open cluster	Cassiopeia	The higher in the sky and the larger the telescope, the better it gets.

INDEX

Acknowledgments: Photographs by page number unless otherwise credited: **9** NOAA; **10, 103, 109, 141, 191** Optical Vision Ltd; **31** International Dark Skies Association; **45, 176l** Martin Lewis; **55, 154** Dave Tyler; **56, 74** David Arditti; **68** Anthony Wesley; **107** Meade; **120, 121** Jamie Cooper; **124** Sally Scagell; **128** Richard Taylor; **162** Damian Peach; **174t, 178** ESA/NASA. All other photographs by the author. Artworks on pages **14, 17, 38, 118, 129, 168, 192, 196** by Jonathan Bell (© Philip's).